Chief Officer

First Edition

VALIDATED BY
THE INTERNATIONAL FIRE SERVICE TRAINING ASSOCIATION

PUBLISHED BY
FIRE PROTECTION PUBLICATIONS
OKLAHOMA STATE UNIVERSITY

Dedication

This manual is dedicated to the members of that unselfish organization of men and women who hold devotion to duty above personal risk, who count sincerity of service above personal comfort and convenience, who strive unceasingly to find better ways of protecting the lives, homes and property of their fellow citizens from the ravages of fire and other disasters . . . **The Firefighters of All Nations.**

Dear Firefighter:

The International Fire Service Training Association (IFSTA) is an organization that exists for the purpose of serving firefighters' training needs. IFSTA is a member of the Joint Council of National Fire Organizations. Fire Protection Publications is the publisher of IFSTA materials. Fire Protection Publications staff members participate in the National Fire Protection Association and the International Society of Fire Service Instructors.

If you need additional information concerning these organizations or assistance with manual orders, contact:

Customer Services
Fire Protection Publications
Oklahoma State University
Stillwater, OK 74078
(800) 654-4055 in Continental United States

For assistance with training materials, recommended material for inclusion in a manual, or questions on manual content, contact:

Technical Services
Fire Protection Publications
Oklahoma State University
Stillwater, OK 74078
(405) 624-5723

First Printing — January 1985
Second Printing — July 1985

Oklahoma State University in compliance with Title VI of the Civil Rights Act of 1964 and Title IX of the Educational Amendments of 1972 (Higher Education Act) does not discriminate on the basis of race, color, national origin or sex in any of its policies, practices or procedures. This provision includes but is not limited to admissions, employment, financial aid and educational services.

© *1984 by the Board of Regents, Oklahoma State University*
All rights reserved
ISBN 0-87939-056-5
Library of Congress 84-73143
First Edition
Printed in the United States of America

Table of Contents

	INTRODUCTION	1
1	**PLANNING AS A MANAGERIAL FUNCTION**	**5**
	Why Planning is Essential	5
	The Chief Officer's Responsibilities in Planning	6
	Elements of the Planning Process	6
	Planning Limitations and Problems	7
	Positive Results From Planning	8
	The Effects of Organizational Size on Planning	9
	The Flowchart as a Planning Tool	11
	County and Regional Cooperation	13
	Outside Mutual Aid	13
	Mutual Aid	13
	Automatic First-Alarm Mutual Aid	15
	Regional Planning for Fire Protection Services	16
	Disaster Preparedness Planning	17
	Planning for the Cascade Effects of Disaster	20
	Special Project Planning	22
	Community Master Fire Planning	26
	Activation and Plan Identification	26
	Problem Identification	26
	Plan Formulation	28
	Plan Implementation	29
	The National Interagency Incident Management System (NIIMS)	29
	Future Planning Activities	31
2	**HIRING PRACTICES**	**35**
	Equal Employment Opportunity and Affirmative Action	37
	Affirmative Action Programs	38
	Management Commitment	38
	The Affirmative Action Plan	39
	Affirmative Action Implementation	41
	Affirmative Action Documentation	41
	Test Validity	43
	Recruitment	45
	Employment Criteria	46
	Civilian Employees in the Fire Department	46
3	**FIRE COMPANY STAFFING**	**51**
	Historical Staffing Trends	51

	Fire Company Staffing Studies	53
	Wisconsin Study	53
	Dallas Study	53
	Staffing Study Conclusions	54
	Factors Affecting Staffing	54
	Company Strength	54
	Response Distance	56
	Work Week and Leave Allowances	57
	Alternative Sources of Company Staffing	57

4 LABOR RELATIONS . . . 63
Development of Public Employee Unions . . . 63
   Reasons for Unionization . . . 65
   Union Organization . . . 66
Contract Negotiations . . . 67
   Maintaining Open Communications . . . 68
The Negotiation Process . . . 69
   Bargaining Preparation . . . 70
   Presenting Proposals . . . 71
   Scheduling Bargaining Sessions . . . 71
   Negotiating Contract Issues . . . 72
Ways to Handle an Impasse . . . 72
Strike Management . . . 74
   Contingency Plans . . . 75
   Anticipating Employee Actions . . . 77
   Public Relations and Communications . . . 78
Internal Procedures . . . 80
   Management Tasks and Responsibilities . . . 81
   Settling the Strike . . . 84
   Final Considerations . . . 86

5 INFORMATION MANAGEMENT . . . 91
Record Systems . . . 93
Report Filing Systems . . . 95
   Computer Storage . . . 96
   Statistical Analysis . . . 96
   Information Applications . . . 99
Information Transfers . . . 100
Public Relations and Information . . . 102
   Public Relations . . . 102
   Public Information . . . 104

6 FIRE DEPARTMENT BUDGETING . . . 111
Types of Budgets . . . 111
Budgeting Methods . . . 113
   The Line-Item Budget . . . 113
   The Performance Budget . . . 115
   The Program Budget . . . 116

       The Planning-Programming-Budgeting System 120
       The Integrative Budgeting System (IBS) 120
       Zero-Based Budgeting (ZBB) 122
   The Budgetary Process 124
   Budget Controls 127
   Alternative Sources of Funds 128
       Service Fees 129
       Public and Private Grants 129
       Self-Insurance 131

7   EMERGENCY MEDICAL SERVICES 135
   EMS and the Fire Service 135
   Determining the Need for EMS 136
   Levels of EMS 139
       First Responder 140
       EMT-Basic 140
       EMT-Intermediate 141
       Paramedic 141
   EMS Certification 143
   Supplemental EMS Assistance 145
   Good Samaritan Laws 145
   Managing EMS 146
       Firefighter/EMTs (Cross-Trained) 146
       Single-Function EMS Personnel 147
       Transportation of Patients 149
   Stress and Burnout 149

8   FIRE COMMUNICATIONS SYSTEMS 153
   Alarm Systems 153
       Public Telephone System 153
       Telegraph Systems 155
       Telephone Fire Alarm Systems 156
       Radio Systems 157
       Private Alarm Systems 161
   Alarm Centers 163
       Staffing 164
       Dispatching Procedures 165
       Dispatching for Emergency Medical Services 167
   Computer-Aided Dispatch 169
   Evaluating the Communications System 172

9   SAFETY IN THE FIRE SERVICE 177
   Safety as an Economic Issue 179
   Documentation 181
   Physical Fitness as a Safety Issue 183

10	THE POLITICAL ARENA	191
	Becoming an Effective Leader on the Local Scene	192
	The Chief as Manager	192
	Increasing the Visibility of the Department	193
	Working to Pass Local Legislation	196
	State Politics	197
	The Legislative Process	197
	The Importance of Unity	199
	Communications Regarding Legislation	199
	Letter Writing	199
	Public Hearings	201
	Lobbying	201
	APPENDIX A	**202**
	APPENDIX B	**204**

List of Tables

1.1	Comparison of Large and Small Fire Departments	9
4.1	Firefighter Strikes 1976-1981	75

THE INTERNATIONAL FIRE SERVICE TRAINING ASSOCIATION

The International Fire Service Training Association is an educational alliance organized to develop training material for the fire service. The annual meeting of its membership consists of a workshop conference which has several objectives —

> . . . to develop training material for publication
> . . . to validate training material for publication
> . . . to check proposed rough drafts for errors
> . . . to add new techniques and developments
> . . . to delete obsolete and outmoded methods
> . . . to upgrade the fire service through training

This training association was formed in November 1934, when the Western Actuarial Bureau sponsored a conference in Kansas City, Missouri, to determine how all agencies that were interested in publishing fire service training material could coordinate their efforts. Four states were represented at this conference and it was decided that, since the representatives from Oklahoma had done some pioneering in fire training manual development, other interested states should join forces with them. This merger made it possible to develop nationally recognized training material which was broader in scope than material published by an individual state agency. This merger further made possible a reduction in publication costs, since it enabled each state to benefit from the economy of relatively large printing orders. These savings would not be possible if each individual state developed and published its own training material.

From the original four states, the adoption list has grown to forty-four American States; six Canadian Provinces; the British Territory of Bermuda; the Australian State of Queensland; the International Civil Aviation Organization Training Centre in Beirut, Lebanon; the Department of National Defence of Canada; the Department of the Army of the United States; the Department of the Navy of the United States; the United States Air Force; the United States Bureau of Indian Affairs; The United States General Services Administration; and the National Aeronautics and Space Administration (NASA). Representatives from the various adopting agencies serve as a voluntary group of individuals who govern policies, recommend procedures, and validate material before it is published. Most of the representatives are members of other international fire protection organizations and this meeting brings together individuals from several related and allied fields, such as:

> . . . key fire department executives and drillmasters,
> . . . educators from colleges and universities,
> . . . representatives from governmental agencies,
> . . . delegates of firefighter associations and organizations, and
> . . . engineers from the fire insurance industry.

This unique feature provides a close relationship between the International Fire Service Training Association and other fire protection agencies, which helps to correlate the efforts of all concerned.

The publications of the International Fire Service Training Association are compatible with the National Fire Protection Association's Standard 1001, "Fire Fighter Professional Qualifications (1981)," and the International Association of Fire Fighters/International Association of Fire Chiefs "National Apprenticeship and Training Standards for the Fire Fighter." The standards are an effort to attain professional status through progressive training. The NFPA and IAFF/IAFC Standards were prepared in cooperation with the Joint Council of National Fire Service Organizations of which IFSTA is a member.

The International Fire Service Training Association meets each July at Oklahoma State University, Stillwater, Oklahoma. Fire Protection Publications at Oklahoma State University publishes all IFSTA training manuals and texts. This department is responsible to the executive board of the association. While most of the IFSTA training manuals can be used for self-instruction, they are best suited to group work under a qualified instructor.

Preface

The primary goal of the fire service has always been to reduce life and property loss due to fire. While the goal remains the same, the fire service has changed: fire departments are being asked to provide a wider range of services at the same time operating budgets and personnel have been cut. Today's chief officer must possess planning, administrative, and political expertise in addition to fire suppression skills. This manual is designed to acquaint chief officers and their subordinates with the management skills needed to plan and maintain an efficient and cost-effective fire department.

Acknowledgements and grateful thanks are extended to the members of the validating committee, who assisted with the final draft of this manual:

Chairman
Clell West
Fire Chief, Las Vegas Fire Department
Las Vegas, Nevada

Vice-Chairman
Ronald L. Callahan
Indiana State Fire Instructors Association
Indianapolis, Indiana

Secretary
Paul H. Boecker
Fire Chief, Lisle-Woodridge Fire District
Lisle, Illinois

Dennis Compton
Asst. Fire Chief, Phoenix Fire Department
Phoenix, Arizona

James McKiernan
National Fire Protection Association
Quincy, Massachusetts

Other persons assisting on the committee during its tenure were

Eugene Kiefer
Insurance Services Office of Oklahoma (retired)

Fred Davis
Fire Chief, Nashville Fire Department

Bobby Mowles
Oklahoma City Fire Department (retired)

Robert M. Porter
Bancroft Fire Protection District (retired)

Leonard Grimstead
Illinois Division of Personnel Standards and Education

Jerry Carter
Fire Chief, Largo Fire Department

K.G. Hulme
Insurance Services Office of Tennessee (retired)

Jim McSwain
Fire Chief, Lawrence Fire Department

Dean Filer
Fire Chief, Central Yavapai Fire District

Eric F. Haussermann
Fire Service Training Specialist
Oklahoma State University

Lee Daughtery
Chief, Littleton Fire Department

Steve Tinberg
Elkhart Brass

William N. (Bill) Jacobs
Asst. Chief, Aurora Fire Department

Lee Hustead
Tennessee Valley Authority

Gratitude is expressed to the following individuals for their expert advice and research efforts:

Al Bauman, U.S. Bureau of Labor Statistics
Colin Campbell, International Association of Fire Chiefs
Ron Coleman, Chief, City of San Clemente, California Fire Department
Carl Holmes, Carl Holmes and Associates
Glenn Pribbenow, Fire Service Training Specialist
Charles Rule, National Fire Protection Association
Pete Stavros, Oklahoma State Firefighters Association

Gratitude is also extended to the following members of the Fire Protection Publications staff who made the final publication of the manual possible:

Lynne C. Murnane	Associate Editor
William J. Vandevort	Associate Editor
Charles Donaldson	Associate Editor
David England	Associate Editor
Mike Buchholz	Publications Editor
Don Davis	Coordinator, Publications Production
Ann Moffat	Graphic Designer
Lynda Halley	Graphic Artist
Karen Murphy	Phototypesetter Operator II
Desa Porter	Phototypesetter Operator II
Carol Smith	Publications Validation Assistant
Cindy Brakhage	Unit Assistant
Gary M. Courtney	Research Technician
Henry R. Moore	Research Technician
Scott A. Stookey	Research Technician

Gene P. Carlson
Editor

COVER PHOTO
Courtesy of *Fire Chief* Magazine (from the July 1982 cover).

Introduction

The job of chief officer has become more complex and demanding than ever. The combination of an ever-increasing fire problem, spiraling personnel and equipment costs, and the development of new technologies and methods for decision making requires far more than expertise in fire suppression. Today's chief officer is expected to coordinate emergency medical services, engage in master planning, plan for possible disasters, analyze cost data, evaluate data processing services, deal with labor unions, and draft legislation. The chief must be manager, administrator, and politician.

Although these skills are relatively new for the chief officer, they are vital. The role of the fire service is changing from solely fire suppression to functioning as part of a fire control and emergency response system that includes fire prevention, public education, emergency medical services, hazardous materials control, and fire suppression. The team approach requires the ability to form and maintain good working relationships with other emergency service personnel, whether at the local, state, or federal level.

Also, the attitude of the general public is changing. In recent years there have been taxpayer revolts and a general cutback in the amount of money available for public service agencies. The fire department is no exception: the public will no longer foot the bill for fire protection and emergency medical services without question. The fire department is now looked on as a service business, and the chief is its executive officer. Since public service agencies are competing for increasingly scarce tax dollars, the fire department is expected to be run in a cost-effective manner on sound business principles. The public is asking: "What is the fire department doing for us?" and the department must be ready to demonstrate, with facts and figures, just how the community benefits from its services.

Whether or not the fire service is changing is no longer a question for debate. The question is whether the fire service will control those changes or will be placed in the position of reacting to changes imposed by others.

SCOPE AND PURPOSE

The **Chief Officer** manual is designed for the chief officer and the firefighter who wishes to become a chief officer. The manual provides an overview of the skills necessary to manage the overall operations of the fire department: planning, hiring practices, staffing, labor relations, information management, budgeting, emergency medical services, communications systems, safety, and political activity. Where appropriate, chapters are referenced to NFPA Standard 1021, *Fire Fighter Professional Qualifications,* for levels V and VI. Appendix B contains references and suggested additional reading for each chapter.

Planning as a Managerial Function

1

NFPA STANDARD 1021
STANDARD FOR FIRE OFFICER
PROFESSIONAL QUALIFICATIONS

Fire Officer VI

7-1 Master Planning. The Fire Officer VI, given rules, codes, operating procedures and objectives designed to provide fire protection for a geographical area, and the current physical plant and facilities, and advanced planning projections for the area, shall:

    (a) analyze trends in urbanization, building construction, population, etc., as they affect the area

    (b) submit recommendations for long range plans which are designed to meet problems inherent in the projected changes in the area

    (c) analyze methods and techniques designed to implement long range programs

    (d) document a need for change based on advanced planning projections

    (e) demonstrate a consideration for the cost-benefit ratio in recommendations submitted.*

7-4 Public Administration

7-4.1 The Fire Officer VI, given present physical and geographical layouts of a simulated area and advanced planning projections for the area, shall:

    (a) evaluate alternatives and select the programs best suited to serve the area in the future

    (b) establish time tables for administration of the programs

    (c) describe the method for implementing the programs and evaluating the results.*

The above NFPA standards are addressed from a general management perspective.

*Reprinted by permission from NFPA No. 1021, *Standard for Fire Officer Professional Qualifications*. Copyright 1983, National Fire Protection Association, Boston, MA.

Chapter 1
Planning as a Managerial Function

WHY PLANNING IS ESSENTIAL

Planning is a managerial function that determines in advance what an organization, a subunit, or an individual should do, and how. It is the foundation of the management process: planning, organizing, directing, and controlling. In fact, planning must be done first to complete the rest of the process.

Fire department managers must deal with projected changes in population and its socioeconomic makeup, inflation, hazardous materials use management, high-rise construction, and other local conditions that affect the fire service. They may have to modify long-range planning yearly because of unforeseen changes. Reductions in revenue sharing will have a variety of effects. Playing on the emotions of elected officials and using the Insurance Services Office (ISO) grading schedule as a crutch to gain budget increases or maintain present funding are tactics that no longer work.

Why plan? Because objective planning will help the department identify issues and set priorities based on resources when internal audits, city hall pressure, or taxpayer demands force the department to change direction. Departments can develop an information base to improve decision-making, and unit managers can identify poor performance. A realistic approach to planning can improve communications, internal coordination, and satisfy training needs. Departments can adjust existing programs and organizational strengths and weaknesses with changes in their own fiscal and social environment. The special needs of the department and its governing jurisdiction might indicate other reasons for planning. In the past, planning efforts were aimed at expansion: more stations, more people, and more programs. Planning in the future will have to emphasize contraction: doing more with present staff or less. Fire chiefs now think they are fortunate if they maintain current funding.

The Chief Officer's Responsibilities in Planning

Chief fire officers are the key to successful planning. Without their active support, the process is doomed to fail. Chiefs must accept the following responsibilities:

- Understanding that planning is a function of top management that cannot be entirely delegated
- Establishing a favorable climate for planning
- Making sure the design of the planning process fits the department
- Being accessible to planning leaders
- Getting involved, when necessary, in evaluation and feedback
- Keeping superiors informed of planning activities

Elements of the Planning Process

There are a number of elements in the planning process:

- Setting goals. Chief officers have to decide the organization's goals and what the department should be doing (Figure 1.1). Planners have to identify their clients' real needs. Is the project or program need real, perceived, or stylish?

Figure 1.1 The chief officer provides leadership by helping the department define its needs, set goals, and outline objectives to reach those goals.

- Evaluating the planning process. What value does the process have for the organization and its employees? Will the process emphasize strategy, brainstorming, open-mindedness, or achievement? Can the organization manage the process if it deviates from the norm?
- Establishing objectives. Establish objectives that will accomplish the goals. Identify short-range, narrow subobjectives.
- Matching goals with department and city philosophy. Are goals consistent with political and economic realities? Are there hidden agendas?
- Establishing policy. Develop a plan of action to guide policy. Carry out objectives within the constraints of the system.
- Planning the organizational structure. Develop a framework that helps people pull together in accordance with strategy, philosophy, and policies.
- Providing personnel. Recruit, select, and develop people qualified to fill positions in the organization.
- Establishing procedures. Determine and prescribe how employees will carry out important and recurrent activities.
- Providing facilities. Provide facilities to carry out the organization's goals.
- Providing the funding. Make sure that funding is available for personnel, facilities, and equipment.
- Setting standards. Establish performance measures that will best enable the organization to achieve long-term goals.
- Establishing management programs and operational plans. Develop programs governing activities and the use of resources that will enable the department to achieve its goals through policies, procedures, and standards. Employees should have the opportunity to realize their personal goals.
- Devising standards of accountability. Develop facts and figures employees can use to follow organizational strategies, policies, procedures, and programs. Measure organizational and personal performance against established plans and standards.

Planning Limitations and Problems

Planning has some pitfalls, but knowing about them ahead of time will minimize the risks.

Introducing formal planning can cause internal resistance. The threat of change once traumatized the traditional fire department. Loyalty to old methods and old standards and fear of the unknown took precedence over planning. Larger organizations are usually more resistant to change. Chief officers must do a selling job and establish the process's credibility in order to overcome this resistance.

A significant planning effort can be expensive. Chief fire officers must know how much time people will devote to planning at the expense of their normal duties, and weigh the costs and benefits objectively (Figure 1.2).

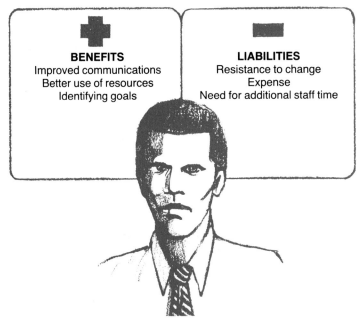

Figure 1.2 Any new program will have a better chance of acceptance if benefits are clearly outlined.

Planning is hard work that can require complex activity with which the organization is unfamiliar. The department might have to call in outside help, which could be costly if talent is not available in-house. And the end result of the process might be unpopular, leaving the department with several unattractive choices.

Finally, planning initiated because of a crisis can end in a quick fix if there is no time for systematic work. The department can be left with short-term results and a long-range disaster.

Positive Results from Planning

The benefits of planning can far outweigh the limitations. Some positive results of the process are listed below.

- Properly changing the organization's direction
- Identifying poor performances
- Identifying issues for management

- Developing better information for decision making
- Concentrating resources according to priorities
- Building a frame of reference for budgeting
- Identifying the strengths and weaknesses of the organization, subunits, and individuals
- Developing improved communication and internal coordination
- Training and developing managers
- Identifying where the organization is now, where it should be, and what programs it needs to achieve its goals
- Setting more realistic goals
- Reviewing and auditing the organization to identify needed change
- Providing a chance to motivate an organization to change the status quo

The Effects of Organizational Size on Planning

Planning is vital regardless of the size of the organization. Chief fire officers should realize that size will affect planning. The following table shows some possible differences between a small and large department (Table 1.1).

TABLE 1.1 Comparison of Large and Small Fire Departments	
LARGE DEPARTMENT	**SMALL DEPARTMENT**
Chief is leader and manages conflict resolutions.	Chief is change motivated.
Important decisions made at top. Routine decisions made by middle managers.	Decisions made at top.
Strata of middle managers insulate chief from firefighters.	Chief and firefighters in close contact.
Authority flows upward/downward by title and not personality.	Chain of command frequently bypassed.
Communications and procedures in writing.	Communication face-to-face and informal.
Multitude of rules/regulations governing subordinate activities.	Loose policies and regulations.
Staff functions more precise.	Staff activities weak.
Formal and impersonal controls established.	Top management involved in employee activity.
Operations detailed and complex.	Operations not complex.
Staff expertise on higher plane.	Minimal commitment to staff support.

An awareness of the way department size affects planning will improve understanding of the planning process and what the department can expect of it. Realize that the factors above might not apply to every department, nor do they imply that large departments plan better.

Departments can use a time-reflex planning system to evaluate the local process of fire protection. The system provides the opportunity to study the proactive and reactive dimensions of the local fire service, a first step in determining how to deliver the correct level of fire protection (Figure 1.3). Collecting time-se-

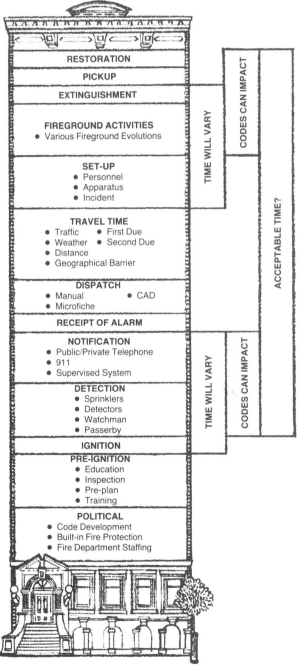

Figure 1.3 Breaking down fire fighting activities into specific steps will allow each to be evaluated to see where improvements can be made.

quence data can force the department to seriously study portions of the response that can be corrected. For example, there is a problem if a paid department takes more than a minute to dispatch apparatus after receiving an alarm.

The system can graphically show lay persons the anatomy of a fire alarm and emphasize that fire fighting does not start with ignition. The U. S. Fire Administration once offered a detailed systems analysis course that helped lay persons evaluate the many facets of the fire service.

When facts might be received with suspicion or even hostility, it is important to have documentation. Unfortunately, fire fighting is not an exact science. For example, the "Comparison of Fire Flow Estimates" chart demonstrates a wide divergence of fire flows for 1,500 square feet (139 m sq) of floor area (Figure 1.4). Estimates range from 100 gpm (378.5 L) to 1,200 gpm (4542.5 L) to control a relatively small fire. Make sure the department can defend the facts it gathers during the planning process.

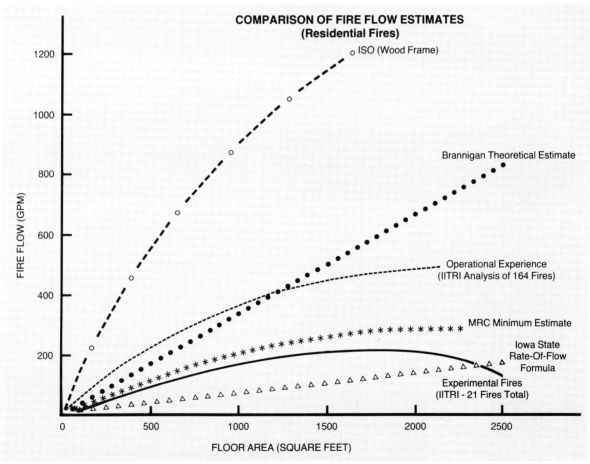

Figure 1.4 Estimates of equipment and supplies needed to fight fires can vary greatly. Experience and careful documentation are necessary for the development of sound plans.

The Flowchart as a Planning Tool

The flowchart is a visual device that can be used to follow a process or program from start to finish. Drawing up a flowchart forces the planner to put the process in a sequence to develop an order for proper completion. Thus, the flowchart is nothing more than a road map of a program.

In the simplest flowchart, the box represents an element of the process. Completing the element satisfactorily moves the user to the next element to be considered. Otherwise, the direction reverses to a point where the user can resolve the problem.

The simple flowchart in Figure 1.5 demonstrates the process of hiring an entry-level employee from the time a vacancy occurs until the rookie completes probationary training. Planners can use many other symbols to produce more complex flowcharts when developing a more sophisticated process.

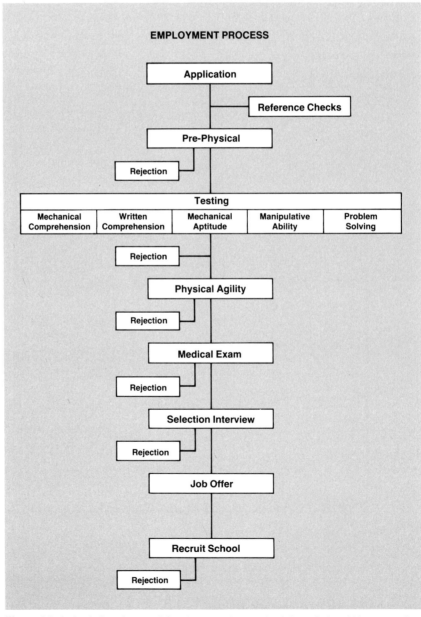

Figure 1.5 A simple flowchart outlining the procedures to be followed when hiring a new firefighter. Each step must be completed successfully before the applicant can move to the next one. *Courtesy of Oklahoma City Fire Department.*

COUNTY AND REGIONAL COOPERATION

Fire protection in America has traditionally been a local responsibility, resulting in a proliferation of fire departments. The fire department in some areas is the catalyst that generates local pride, and being a firefighter gives a person social status. This community pride is admirable, but the department's effectiveness should be judged by its ability to provide services. Some of these functions can be handled better at the county or regional level.

It is easy to say that some county or regional operations are more operationally efficient, cost effective, and logically sound. But it is not easy to get several political jurisdictions together and reduce their local control and autonomy. There are also different levels of regional cooperation:

- Complete isolation: the department feels it neither wants nor needs help.
- Traditional mutual aid: the department asks for help after it arrives at the scene.
- Automatic first-alarm mutual aid: the closest unit responds automatically.
- Central communications.
- Joint purchase of equipment and apparatus.
- Joint recruitment.
- Operational consolidation.
- Full political consolidation.

Outside Aid

Outside aid commonly involves an agreement to provide services to an unprotected area or a smaller fire department for a fee. A township, village, or small suburb can contract with an urban fire department for a prearranged fee, which could be a retainer plus an hourly rate or an hourly per-unit rate.

Milwaukee, for example, has provided outside aid to suburban departments that could not meet the city's mutual aid standards. Milwaukee charged $2,000 per hour per unit or any part thereof, which resulted in a first-alarm hourly assignment fee of $8,150 for three engines, a truck, and a chief officer. Some thought the fee excessive, but it forced some jurisdictions to update local services. Local jurisdictions should be governed by their own conditions and rules.

Mutual Aid

Mutual aid involves the exchange of services without charging fees. One department sends aid at another department's request after the incident commander makes an on-scene assessment (Figure 1.6). Mutual aid provides backup, but depending on when it is called in and the response time, it can be too little, too late. All participating jurisdictions should sign mutual aid agreements, identifying areas of authority, responsibility, and the fiscal obligations (if any) incurred by the jurisdiction in which the incident occurs. Will each jurisdiction pay its own expenses for such items as workmen's compensation, special extinguishing agents, and damaged or lost equipment? The agreement should resolve such concerns before any incident. Also, differences in

EQUIPMENT AVAILABLE FOR MUTUAL ASSISTANCE

FIRE DEPT.	PUMP GPM CAPACITY	2½" HOSE AMOUNT & THREAD	3" AND LARGER HOSE AMOUNT & THREAD	TANKERS AMOUNT OF WATER	GROUND LADDER	AERIAL DEVICES	HEAVY STREAM APPLIANCES
LAKE BLUFF FIRE DEPT.	1 - 1000	800 ft. NST	3 in. - 50 ft. NST 4 in. - 2000 ft. Storz	1 - 1000 gal.	1 - 24 ft. 1 - attic 1 - roof		deluge set
FORT SHERIDAN FIRE DEPT.	1 - 750 1 - 500	2600 ft. NST	140 ft. NST		2 - 35 ft. 2 - attic 2 - roof		
BARRINGTON FIRE DEPT.	2 - 1500 1 - 250	1000 ft. NST	3 in. - 1550 ft. NST	1 - 3000 gal.	1 - 14 ft. folding 1 - 24 ft. 1 - 30 ft. 1 - 35 ft. 1 - 40 ft. 3 - roof	1 - 85 ft. aerial ladder	1 - 500 gpm 1 - 1000 gpm
COUNTRYSIDE FIRE DEPT.	1 - 1000	1000 ft. NST	3 in. - 900 ft. NST	1 - 1000 gal.	1 - 14 ft. 1 - 24 ft.		1 - aerial water gun mounted on pumper
FOX LAKE FIRE DEPT.	1 - 1500 1 - 1000 1 - 350	1550 ft. NST	3 in. - 2100 ft. NST	1 - 3700 gal.	1 - 16 ft. 3 - 24 ft. 1 - 35 ft. 1 - 40 ft. 2 - roof 2 - attic	1 - 65 ft. aerial tower	2 deluge sets
HIGHWOOD FIRE DEPT.	1 - 750 1 - 1000	2600 ft. NST	3 in. - 950 ft. NST		2 - 14 ft. 2 - 35 ft. 1 - 50 ft.	1 - 90 ft. aerial platform	1 deluge gun
KNOLLWOOD FIRE DEPT.	2 - 1000 1 - 650	950 ft. NST	3 in. - 600 ft. NST 4 in. - 1200 ft. Storz	1 - 1000 gal. 1 - 1500 gal. port.	1 - 24 ft.		2 - 500 gpm pipes
LAKE VILLA FIRE DEPT.	1 - 1000 1 - 750	1500 ft. NST		1 - 700 gal. 1 - 1500 gal. port.	2 - 35 ft. 2 - roof 2 - attic		1 - 500 gpm deluge gun mounted on a 4WD truck
NORTH CHICAGO FIRE DEPT.	1 - 1000 1 - 1250	800 ft. NST	3 in. - 1300 ft. NST		1 - 14 ft. 1 - 24 ft. 1 - 35 ft.	1 - 85 ft. aerial tower	1 - 500 gpm 1 - 750 gpm

Figure 1.6 Mutual aid can save time and money by making the most efficient use of neighboring communities' resources. *Adapted from information supplied by Lake County, Illinois Firemen's Association.*

fireground tactics and terminology can lessen the effectiveness of mutual aid. Planners need to take the extra step and include some joint training so that different departments will be able to work together efficiently.

Automatic First-Alarm Mutual Aid

This involves an agreement to provide first-alarm service regardless of political boundaries. The units closest to the incident respond and come under the supervision of the jurisdiction in

FOAM EQUIPMENT	PORTABLE PUMPS GPM	LIGHTING EQUIPMENT	SCBA	AMBULANCES	SPECIAL EQUIPMENT
	1 - elec.-250	2 - 3500 wt. gen. 5 - floodlights	6 - 30 min. MSA	1 - rescue squad	Rear tank dump, generator
95 gal. AFFF 125 lbs. dry foam mix	2 - 300	1 - 2500 wt. gen.	6 - 30 min. MSA		K-12, Porta-Power, cribbing, explosimeter, gas generator, smoke ejector, brush equip., air crash rescue gear, proximity suits
2 - pickup units 100 gal. foam in 5 gal. containers	1 - 100	1 - 5000 wt. port. gen.	6 - 30 min. Scotts & spare bottles	1 - MICU 2 - patient	Acetylene torch, air bottles, air cascade sys., smoke ejector, air chisel, K-12, Porta-Powers and come-alongs, salvage equip., 2 Hurst tools (24 and 32), low level strainer, siphon device, 10 inch rear dump, 4WD brush unit
	2 - 250	1 - 1500 wt. gen.	6 - 30 min. Survivair	1 - 2 patient	Smoke ejector, water vac, salvage covers, Jeep with small pump, boat
40 gal. foam eductor 1 - foam unit	1 - 300	1 - 4500 wt. port. gen. 2 - 16 ft. light towers w/ 6- 5000 wt. bulbs on each 10 kw gen.	6 - 30 min. MSA 6 - 30 min. Scotts	2 - rescue squad	2 smoke ejectors, K-12 saw, floating strain, 2 back packs in basket, Manpower Squad and Divers Van, 3 boats, air bank, Hurst tool, hydraulic saw, elec. chisel, elec. winch, air bags, 2 FWD Jeeps w/100 gal. tanks, Stokes basket, 4WD minipumper brush truck
4 - 5 gal. cans 2 - foam applicators		1 - 3000 wt. gen. 1 - 2000 wt. gen.	4 - 30 min. Scotts	1 - 4 patient	K-12 saw, Porta-Power
high-expansion foam gen. Angus foam inductor Foamaster nozzle 60 GPM 15 gal. AFFF 500 gal. deluge		1 - 3000 wt. gen. lighting equip.	2 - 30 min. Survivair		
foam eductor 1 - foam proportioner	1 - 250	1 - 6500 wt. gen. 2 - floodlights	3 - 30 min. Scotts & spare bottles	1 - rescue squad	Generator, smoke ejector, Hurst tool, Porta-Power, K-12 saw, air bank, stretcher, 2 Jeeps with 150 gal. tanks
	1 - 500 1 - 350	1 - 5000 wt. gen. 2 - floodlights	6 - 30 min. Scotts	1 - 4 patient 1 - rescue squad	Hurst tool, chain saw, Porta-Power, air chisel, K-12 saw

which the incident occurs. The system produces gains in productivity, quicker response, and expansion of familiar and available resources in emergencies. Fire departments in northern Virginia have successfully used first-alarm mutual aid for more than five years with substantial savings in annual operating funds and capital outlay for new fire stations.

REGIONAL PLANNING FOR FIRE PROTECTION SERVICES

Planning for regional fire service is controversial and difficult. Obstacles facing regional planners include protection of turf, fear of the unknown, loss of power, and petty jealousy. They have to study geographic and fiscal boundaries in depth. They must develop formulas for fair-share payments by each jurisdiction. But it can be done. The Glendale-Burbank area of California and Hamilton County (Cincinnati), Ohio, for example, have both developed cost-sharing agreements for joint communications centers.

There are several advantages in drawing up interjurisdictional joint-power agreements:

- They fill gaps in expertise.
- They allow better use of equipment and facilities.
- They make possible a higher quality of service.
- They can save money.

Departments can use contracts and joint-service agreements to set up functional consolidation. Figure 1.7 shows a simple planning sequence to use.

The process will generate the following questions:

- What is the goal of the agreement?
- Is it legal to enter into such agreements?
- What are the total short-term and long-term costs?
- What factors will the department use to determine costs? Area? Population? Number of incidents?

Figure 1.7 A carefully thought-out plan can eliminate many problems when establishing regional services.

DISASTER PREPAREDNESS PLANNING

An emergency is a sudden, unexpected occurrence or set of circumstances that demands immediate action. The way an emergency is handled may determine whether or not severe consequences will follow. Disaster planning is an attempt, prior to the actual crisis, to determine emergency demands in order to make the community response more effective. Disaster planning involves four categories of elements or activities:

Mitigation: Those elements that can prevent or minimize the impact of a catastrophe before the incident occurs — zoning, building and fire prevention codes, insurance, and public education and information.

Preparedness: Those functions that can prepare a community to cope with disaster — installing warning systems, maintaining inventories, developing operational plans and training in their use, realistically evaluating mutual aid plans.

Response: Those activities designed to manage an incident — implementing plans for warning and evacuating residents of the disaster area; providing them food, shelter, and medical treatment; putting out fires; and conducting search and rescue operations.

Recovery: Those activities performed to restore the community — making emergency repairs, abating further disasters, setting up temporary housing, restoring utilities, and reconstructing the damaged or destroyed portion of the city.

The four categories are definitely related and should be considered a system.

> Since rescue and relief activities must be well coordinated to be effective, the disaster plan should be tested by holding disaster drills. Then the plan can be evaluated to see where improvements in communications, training, and medical care can be improved. Disaster plans should be reviewed and updated annually.

Unfortunately, disaster planning does not get enough attention in many governmental agencies until after a disaster has occurred. The problems of managing day-to-day operations take precedence over realistic disaster planning. The negative attitude of some local government officials, who say that "it can't happen here," that "we can handle anything," further impedes the process. The fire service must provide leadership to develop a pragmatic plan that can cope with disaster. Otherwise, the organization will be overwhelmed in the early stages of a large incident.

When developing a plan, an evaluation of the jurisdiction's hazard potential and the determination of priorities for action

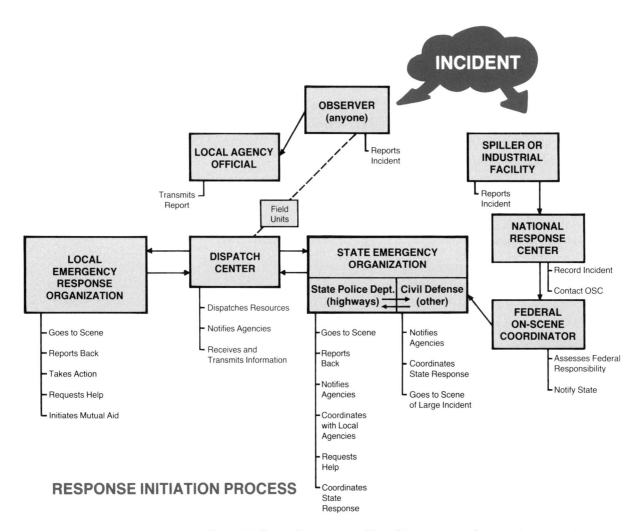

Figure 1.8 Once a disaster occurs, it is too late to try to coordinate services among local, state, and federal agencies. Each department must know beforehand what is needed to mitigate the disaster.

must be made (Figure 1.8). Special attention should be given to hazards that could lead to major emergencies or disasters, while routine problems should receive a lower priority. Consideration should also be given to probability, history, vulnerability, risk, and maximum threat. A cross section of disciplines should be used to ensure objectivity. Separate criteria should be developed for each incident and the results integrated into the overall plan.

Potential hazards fall into the following categories:

- Transportation: rail, air, road, waterway, pipeline
- Bad weather: snow, ice, drought, wind, hail
- Nuclear incidents: war, power plant accident, waste spill
- Natural disasters: earthquake, hurricane or typhoon, tornado, flood, tsunami
- Utility failures: gas, electrical, water, telephone
- Industrial incident: spill, leak, explosion

- Fire: wildland, high-rise, institution, conflagration
- Structural collapse

The Association of Bay Area Governments (ABAG) of the San Francisco Bay region has developed a series of questions to help with disaster planning. Below is a modified version.

1. What causes the disaster?

Disasters are caused by natural phenomena, such as abnormal weather, or by human failure, such as a rail accident. An air pollution problem is a combination of both. Knowing the cause can aid in reducing the hazard or dealing with it more efficiently.

2. How does damage occur?

Injury and death are caused by the sudden release of energy, poisoning (chemicals, contact with hazardous material and so on) or drastic changes in the environment. Property damage occurs when the norm changes.

3. Are there seasonal or other time constraints?

Natural disasters are often associated with the seasons. Severe cold is associated with winter. Spring spawns tornadoes. A dry summer will generate wildland fires. Hurricanes appear in the fall. The time of day can also have an effect on how departments handle disaster. A school fire at 3 p.m. is a different tactical and strategic challenge than the same fire at 3 a.m. Departments that consider the effects of the seasons and the time of day can prepare more efficiently for particular incidents, but planners should not be lulled into thinking that disaster is a prisoner of time and the seasons.

4. Are there geographical constraints?

Some hazards are linked to geographic locations or land arrangement. Tidal waves are associated with coastal areas, landslides with weak land formations or lack of stable vegetation. Rail incidents occur within defined areas and aircraft crashes are more likely within several miles of an airport.

5. What is the geographical extent of the disaster, and where is the effect most intense?

A snow disaster could affect many square miles while a landslide might cover a few acres. The effects of the snowstorm could be relatively uniform throughout the area but more severe in isolated areas. There could be serious damage in some areas and none in others. Air pollution emergencies tend to be uniform over an area, while tornado damage can be haphazard.

6. Can an imminent disaster be forecast?

Some agencies can forecast disaster and issue advisories if they know and heed the conditions. A heavy snowstorm and unseasonably warm weather can signal an impending flood. Falling barometric pressures, high winds, and prolonged periods of high temperatures and low humidity are the ingredients of wildland fires. Locked hotel exits or transportation of hazardous materials through a downtown area are obvious ingredients for disaster.

7. Is there a warning period? If so, how long?

Warning is critical to protect life and property. The amount of warning time often determines the protective measures a department can take. Devices such as smoke detectors, automatic fire sprinklers, and warning sirens can alert anyone from individual building occupants to large groups of residents in intermetropolitan areas. But remember that for some disasters, such as airplane crashes or train wrecks, there is little or no warning.

8. How quickly can disaster strike?

Tornadoes, earthquakes, transportation incidents, and similar disasters can strike suddenly and with great intensity. Others, such as snowstorms or droughts, can gradually build to a peak.

9. Is the intensity of a disaster predictable?

Experts might be able to predict the intensity of a flood or a power failure, but predicting the intensity of a nuclear war or an epidemic is nearly impossible.

10. How long will the impact last?

Tornadoes can appear and disappear quickly, while a three-day snowstorm gives the department more time to mobilize resources and assess the magnitude of the incident. An epidemic can develop slowly, giving officials a chance to reduce the effects.

Planning for the Cascade Effects of Disaster

Some disasters trigger a multitude of secondary incidents that can create a more intense disaster management problem than the primary event. For example, Figure 1.9 illustrates the secondary disasters that could be caused by an earthquake.

Planners must consider these secondary disasters, which could be of major consequence. The plan should provide ways to manage these secondary incidents, which might have to be handled simultaneously with the effects of the primary incident.

Disaster planners make a common mistake when they assign a number of functions to a single agency, such as the local fire de-

Planning as a Managerial Function **21**

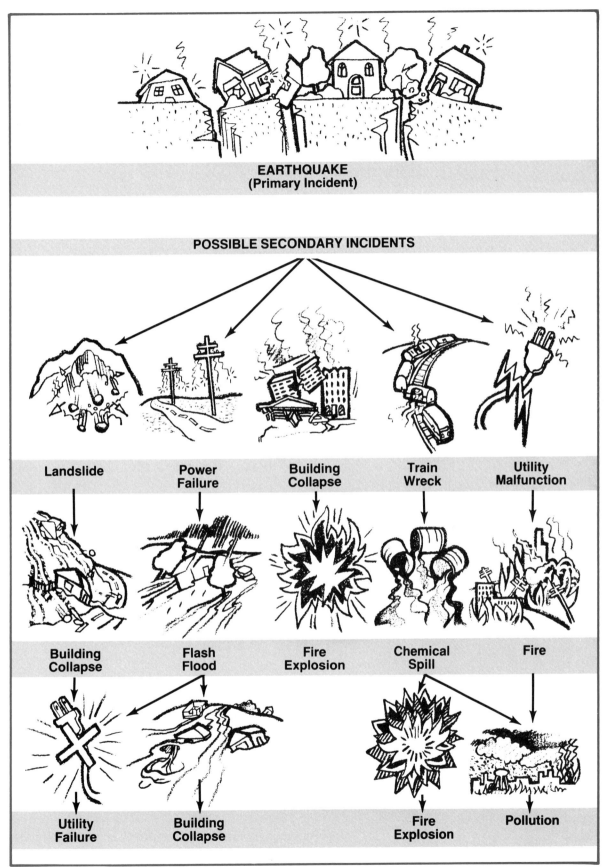

Figure 1.9 The repercussions from a primary incident can be as serious as the incident itself. Fire departments will find their resources stretched to the limits as they are called on to assist with fires, chemical spills, and rescues.

partment, without providing for mutual aid in case of a large disaster. Disaster plans typically give the fire department responsibility for warning, fire suppression, rescue, emergency medical services, triage, radiological monitoring, evacuation, and so on. And all this is to be managed with only 10 firefighters on duty, a totally irrational approach if there are several secondary incidents generated from the primary incident.

Pragmatic planners will identify these deficiencies and provide for alternative resources.

SPECIAL PROJECT PLANNING

In addition to the normal disaster plans for extreme weather conditions, departments must develop special project plans, such as disaster plans for hazardous material incidents (Figure 1.10). Montgomery County, Md., has developed such a planning guide. But before developing the plan, the county set the following objectives:

- Determine ahead of time the types of hazardous material incidents that can occur in the area.
- Determine a safe and orderly process for dealing with each incident.
- Give top priority to protecting life.
- Restore services as soon as possible after abating the incident.

The hazardous materials disaster plan outline provides the framework for planning. It should include the following elements:

A. Introduction
- Reason for developing the plan
- Purpose
- Scope
- Distribution

B. Legal Authority: laws and procedures under which the government can act at an incident

C. Plan Development: levels by which to expand a plan to handle more complex incidents

Level I: To cover a major hazardous materials incident that can be managed with local resources

Level II: To cover an extreme emergency that requires help from outside agencies

HAZARDOUS MATERIALS DISASTER CHECKLIST

Pre-Disaster Period

1. Conduct hazard analysis
   a. Transportation modes
      1. Air
      2. Rail
      3. Water
      4. Pipeline
   b. Fixed facilities
      1. Factories
      2. Bulk storage
      3. Shipping and transfer
2. Conduct a survey of hazardous materials:
   a. Location
   b. Type
   c. Quantity
3. Evaluate resources available for dealing with specific locations and classes (flammable gases, corrosives, etc.) of hazardous materials.
4. Refer to transportation checklist for suggested surveys of transportation modes.
5. Determine deficiencies in individual department operating procedures.
6. Develop hazardous materials training program.
7. Develop standard operating procedures for hazardous materials incidents.
8. Establish mutual aid agreements with industries for:
   a. special fire protection
   b. suppression agents
   c. special container patch kits
   d. technical experts
   e. spill control and clean-up equipment personnel
9. Develop pre-emergency response plans for potential transportation incidents and fixed facilities. Consider:
   a. quantity of hazardous material involved
   b. health problems
   c. fire danger
   d. reactivity with suppression agents
   e. potential dispersion areas
   f. life and property and environmental exposures
   g. control/shutoff valve locations
   h. special equipment required

Disaster Period

1. Determine the presence of hazardous materials. Consider:
   a. transportation vehicles
   b. dumps/waste sites
   c. construction areas
   d. fixed facilities
2. Estimate the potential harm to life, property and the environment. Consider:
   a. container size
   b. shape
   c. pressure
   d. quantity
3. Choose a response objective and consider options. Protect life exposure as necessary. Consider:
   a. Intervention for immediate life-threatening rescue, if required.
   b. Withdrawal from area for identification of material involved and further assessment.
   c. Total withdrawal and evacuation to an estimated safe area.
4. Identify the material involved. Look for:
   a. Use levels
      1. Industrial sites (higher concentration potential)
      2. Home use (weak concentrations)
   b. Containers
      1. Sizes
      2. Shapes
      3. Configurations
   c. Container marking systems
      1. Special color codings
      2. Placards/labels (D.O.T., NFPA 704, United Nations Number)
      3. Stenciled tank identification numbers
      4. Company signs, product names
   d. Documents
      1. Waybill
      2. Consist
      3. Invoices
      4. Supply/Stock Inventory Lists
5. Contact manufacturer, shipper, etc., as required by contacting the Chemical Transportation Emergency Center (CHEMTREC 800-424-9300).
6. Re-evaluate emergency with new information.
7. Monitor progress throughout the incident.

Post-Disaster Period

1. Conduct medical evaluation of personnel as necessary.
2. Evaluate resources, inventory supplies, equipment damage.
3. Criticize emergency operations.
4. Review, revise, and update standard operating procedures, disaster plan mutual aid agreements as required.
5. Implement training program for correcting deficiencies.

Figure 1.10 A hazardous materials checklist allows planners to evaluate present resources, determine which materials and services will be needed, and outline procedures to follow should a disaster occur. From *Disaster Planning Guidelines For Fire Chiefs*.

Level III: To cover an extreme emergency that requires help from the federal government for more than 24 hours

D. Personnel Notification Procedures
- Fire service
- Law enforcement agencies
- Emergency medical services
- Civil defense

E. Public Notification Procedures
- Radio
- Television
- Newspapers
- Standard signal alert

F. Operational Authority
- Chain of command for each level of the incident
- Command post with an officer for communication, logistics, safety, equipment, decontamination, and public information

G. Evacuation Plan
- Responsibility of government agencies
- Operation of evacuation centers, including location, opening, public notification, evacuee records, staffing, communications, and logistics of food, clothing, beds, and rest rooms
- Care of animals
- Medical care
- Reunion of families
- Traffic control (handling congestion during evacuation, securing access roads after evacuation, and dealing with breakdowns when transporting evacuees)
- Evacuation of special facilities (hospitals, nursing homes, hotels or motels, and public assembly facilities such as theaters, sports arenas, and shopping centers)
- Evacuation of the handicapped (hearing- and sight-impaired, nondrivers, and those without vehicles)
- Communication through the media for foreign-speaking residents and evacuees
- Information for those returning to the evacuated area

H. Organization Resource List for Personnel and Equipment
   - Government agencies (fire, law enforcement, civil defense, emergency medical services, armed forces, environment, health, recreation, transportation, education, and utilities such as gas, water, sewer, and electric)
   - Private agencies (hospitals, contractors, building supply firms, spill control companies, chemical companies, transportation companies, and service organizations such as the Red Cross)

I. Maps
   - Street
   - Topographical
   - Medical facilities
   - Evacuation centers
   - Transportation routes
   - Evacuation routes
   - Pipeline and gas mains
   - Electrical lines
   - Water lines
   - Sanitary and storm sewers

J. Technical Assistance Sources
   - Government agencies
   - Industry groups
   - Educational institutions
   - Weather services
   - Reference texts

K. Logistics Plan for Emergency Personnel
   - Food
   - Housing
   - Relief by other on-duty personnel
   - Electricity and telephone service
   - Rest rooms
   - Expendable supplies (gasoline, diesel fuel, oil)
   - Apparatus repair

L. Emergency Medical Services Plan
   - Field evaluation procedures

- Transportation procedures
- Specialized medical facilities
- Evacuation center treatment

M. Revision

The plan must be useful, simple, and easy to read. An unclear plan will not be used during an incident. It should be a guide and reference source. It should also consider the state of a hazardous material: Is it being manufactured, stored, transported, used, or disposed of? Incidents involving hazardous materials in different states might require different approaches.

Planning for transporting hazardous materials is the most difficult because of accessibility, water supply, evacuation, traffic movement, and the unknown quantity and quality of the material.

The plan is worthless unless the department trains its personnel to use it. The plan must also be kept current through constant review.

COMMUNITY MASTER FIRE PLANNING

The pioneer in master planning in the late 1960s and early 1970s was Mountain View, Calif., which created the National Fire Prevention and Control Administration (NFPCA). One of its top priorities was developing a community fire protection master planning process, which many jurisdictions ranging from urban centers to rural communities have tried to adopt. Several fire service people have reported that gathering numerous disciplines into the planning process is one of the major benefits of master planning (Figure 1.11).

The community fire protection planning process includes several elements:

Activation and Plan Identification

Who will be involved and to what extent? Who will be the project manager? What are the functions of the participants and what level of resources will they commit? What is the scope of the plan? How will it be activated? What are its major goals and its anticipated results? What should the planning process and the work schedule be?

Problem Identification

Make an inventory of existing conditions, including the following:

- Fire risks
- Fire protection boundaries, zones, and so on

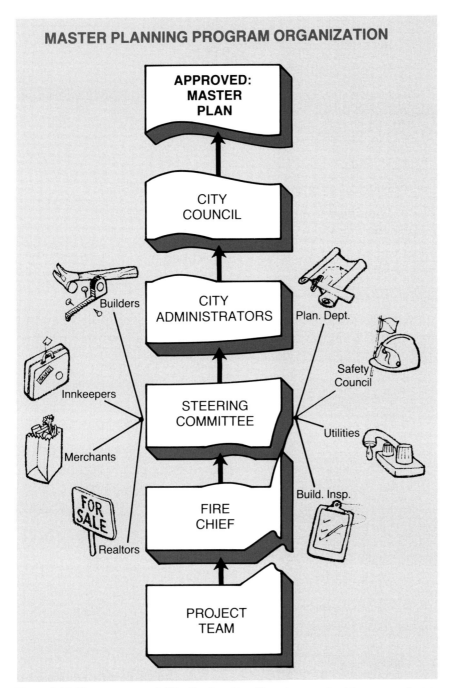

Figure 1.11 The concerns and difficulties faced by different segments of the community are brought out during master planning.

- Physical layout such as water distribution systems, road networks, and natural barriers
- Citizen attitudes toward fire protection and prevention

Examine fire and fire protection trends and make an inventory of existing fire protection resources and facilities:

- Buildings
- Apparatus and equipment
- Personnel (their number and competency)

- Laws, codes, regulations, and programs (their completeness and effectiveness)
- Revenue sources for fire protection

Evaluate programs, plans, and policies that will affect the management of possible fire problems:

- The high-value district
- Density
- Natural or man-made barriers to fire spread
- High-rise or other specialized risks

Develop the trends, existing conditions, and projections into a comprehensive statement of the problem.

Plan Formulation

Appoint planning advisory committees representative of the community to review the literature and policies dealing with fire protection standards. Consider the following:

- The Insurance Services Office (ISO) fire suppression rating schedule
- Recognized building codes that can be developed locally if adopted as minimum standards
- City planning codes dealing with site plans, building setbacks, street width, through streets, and site use
- Model sprinkler codes

Re-examine collected data and develop it into a useful data base. Consider information about the following:

- High-risk areas
- Facility systems
- Demographic history (socioeconomic information)

Develop goals and objectives that will mesh with those of other departments and agencies, such as the public works department, the planning department, and the school system.

Program Development

Using a systems approach, develop specific plans and programs to meet objectives generated earlier in the process. The system should combine fire suppression, education, prevention, and enforcement. Identify new locations for fire stations and levels for staffing and service. (Public Technology, Inc. of Washington, D. C., has programs available.) Departments should

- Determine the programs' fiscal impact.
- Identify alternate programs and assess their fiscal impact.

- Identify legislative requirements and possible constraints.
- Examine the potential private, state, and national resources available.

Plan Implementation

Implementing the plan involves the following:

- Objectively evaluating alternatives and different strategies for implementing them.
- Putting alternatives and strategies in priority order.
- Recommending the final master plan to policymakers after developing a program to sell the recommendations. The latter is especially needed if the plan is controversial and the master plan committee disagreed on it. City policy may limit how much selling government members of the committee can do. Private citizens might be better for the job.
- Setting dates to review the implementation process, make changes, and update the plan.

The U. S. Fire Administration in Emmitsburg, Maryland, has information on implementing plans.

THE NATIONAL INTERAGENCY INCIDENT MANAGEMENT SYSTEM (NIIMS)

There are numerous fire organizations throughout the United States that are dedicated to fire suppression. They usually function well in individual and combined efforts, but this is not always the case. Cooperation in large-scale operations can fail because of differences in terminology, a lack of standards, incompatible apparatus and equipment, a lack of common radio frequencies, and turf disputes.

Federal wildland fire protection agencies and some state forestry agencies have necessarily operated under the Large Fire Organization System (LFOS). LFOS when combined with the National Interagency Fire Qualification System (NIFQS), developed by the National Wildfire Coordinating Group (NWCG), is a complete fire management system. But the system's structure and terminology hindered cooperation between fire departments and other emergency organizations.

The Incident Command System (ICS), which provides a common emergency management structure for agencies that must work together in a crisis, was developed in Southern California. When NWCG realized that someone was developing a parallel system, it initiated a study to analyze existing fire suppression systems and pull the best parts into a single system that could be accepted nationwide. The result was the National Interagency Incident Management System (NIIMS), which consists of five major subsystems (Figure 1.12).

National Interagency Incident Management System NIIMS				
INCIDENT COMMAND SYSTEM	TRAINING	QUALIFICATIONS and CERTIFICATION	PUBLICATIONS MANAGEMENT	SUPPORTING TECHNOLOGY

Figure 1.12 The five segments of the National Interagency Incident Management System (NIIMS) are designed to form a national emergency operations system.

Incident Command System — An on-scene management structure that includes operating requirements, interactive management components, and an organizational and operational structure.

Training — A subsystem that develops and delivers courses to support the ICS organizational and operational structure and other related courses.

Qualifications and Certification — A subsystem that fosters national qualification and certification standards for wildland fire fighting and could eventually foster standards for fire fighting and other emergency public services. These standards could include training, experience, and physical fitness.

Publication Management — A subsystem for developing materials and controlling and distributing publications.

Supporting Technology — A subsystem that could integrate with NIIMS some of the following elements:

- Orthophoto mapping
- Formal decision-making processes
- Multi-agency coordination
- Automated fire modeling
- Infrared technology
- The vegetative fuels data base
- Communications planning

Departments throughout the nation will have to accept NIIMS for the system to function properly. Developers hope the following characteristics will ensure the system's acceptance:

- It contains common standards.
- It applies to any emergency from the smallest to the most complex.
- It can be expanded logically and smoothly.
- It lets agencies keep their autonomy.

- It applies to all fire protection agencies.
- It stresses total mobility and use of the nearest forces.
- It adapts to new technology.
- It allows for previously developed personnel qualifications and certification.
- It lets agencies develop qualifications and certifications that meet regional and local needs.
- It ensures low-cost training, operations, and maintenance.

NIIMS has the potential to benefit emergency services at all levels in the following ways:

- It will provide a fully qualified fire protection agency more quickly by using firefighters and equipment from local, state, and federal agencies.
- It will provide a way to get more help from throughout the United States.
- It makes more effective use of special skills at fires and other emergencies. Groups more experienced in wildland fire fighting could work mostly to control remote fires spreading in vegetation. Local fire departments can work to save life and property and to control wildland fires in more accessible areas. Wildland firefighters could manage structural fires in remote forest areas.

The system could decrease damage to natural resources and property and decrease the time and cost of suppression.

More information about the system is available from FIRE-TIP, Cooperative Fire Protection, Boise Interagency Fire Center, Boise, Idaho.

FUTURE PLANNING ACTIVITIES

Future fire service planners and researchers will have to face some basic and pragmatic issues that, for the most part, are "untouchable" today. Cutback management will force elected officials, city administrators, and chief fire officers to face decisions they are ill equipped to manage now.

For example, the fire service has failed to identify the number of people who should be assigned to engine and truck companies to do the best job of putting out fires. The number across the country runs from one to five, although the latter is rare. What is the best staffing? How many fire companies does a community need, and by what criteria? If the budget permits on-duty staffing of 48, what configuration is best? Twelve crews of four, eight crews of six, eleven crews of four plus a personnel squad of four, or some other variation?

What impact does community structure, density, socioeconomic characteristics, and occupancy have on the number of companies and company staffing? Do different fire risks and different times of the day require different staffing levels?

What is the correlation between response time and staffing levels? Should the first-due company arrive with a crew of two in two minutes or with five in five minutes? Is a highly skilled company of three more effective than a mediocre crew of five?

What effect does built-in fire protection have on staffing patterns? Could the department reduce staffing levels if every structure of more than 5,000 square feet (465 m sq) of floor area had smoke detectors in every sleeping room and built-in sprinklers?

How should the department handle personnel shortages? For economic reasons, some cities supplement the paid force with volunteers. But that can be an emotional decision, accompanied by unfounded charges and countercharges.

Should the department sign a mutual aid pact? It is a question any serious planning effort must consider. Planners have to evaluate the quality, quantity, and availability of mutual aid. In dense urban areas with multiple jurisdictions, first-alarm mutual aid can save millions of dollars in personnel and capital costs annually. But elected officials and operational personnel have to be committed to mutual aid for it to function in the best interest of competent fire protection.

How should departments handle fire insurance? Some departments, chiefly on the West Coast, have looked into self-insurance. Under this system a department could generate revenue and use profits to achieve a greater impact on the local fire problem.

There are three basic dimensions that fire protection planners and decision makers must assess to decide the right level of fire protection for their communities: the risks to life, the risks to property, and the community's ability to pay for the service.

There is no lack of questions for the fire service to answer. If the fire service shows its responsibility by planning soundly and rationally, the answers will be found inside the profession. Otherwise, "failing to plan is planning to fail."

Hiring Practices

2

CHAPTER 2 OBJECTIVES

1. Develop general and specific job descriptions for firefighter applicants.

2. Analyze written, oral, and physical tests given to firefighter applicants to determine whether they are job related.

3. Demonstrate knowledge of equal employment acts and orders that apply to the jurisdiction by developing an affirmative action plan for recruiting firefighter candidates.

Chapter 2
Hiring Practices

A central function of fire department administration is hiring firefighters, training them, and keeping them ready to respond to emergencies. Firefighters are the department's most important resource, and hiring, firing, promoting, and transferring firefighters are among the chief officer's major responsibilities. Personnel management has always been demanding, but it has become even more so since passage of the 1964 Civil Rights Act and subsequent equal employment acts and executive orders. Almost 20 years have passed and controversy still surrounds fire department hiring and firing. Today, fire departments face court-ordered hiring practices because administrators have failed to follow the intent of these laws.

What exactly is required by these civil rights/equal opportunity acts and orders, and what do they mean to the fire department administrator? *Every individual in the United States has the right to seek employment without fearing or experiencing discrimination or unequal pay because of*

- Race
- National origin
- Creed
- Religion
- Handicaps, real or perceived
- Disability
- Sex
- Age

The law, however, does not force employers to hire people physically or mentally incapable of doing the job. Nor does the law prohibit testing to fill positions, but the administrative officer should ensure that the test evaluates the physical and mental capabilities necessary for the position.

A few fire department administrators view equal opportunity legislation and affirmative action programs as government acts forcing fire departments to lower standards and efficiency, but others disagree.

The crux of the issue is this: job placement must be directly associated with the applicant's ability to perform the job (Figure 2.1). Any other criteria are invalid in the eyes of the law. Discrimination is any hiring practice based on criteria other than the applicant's ability to do the job.

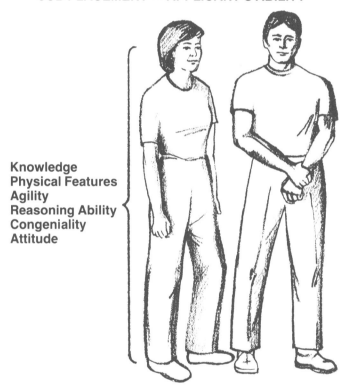

Figure 2.1 It is discriminatory to evaluate an applicant using any other criteria than ability to perform the duties of the job.

Officers should become familiar with specific acts and orders that apply in their jurisdiction. Just because the governing jurisdiction has an affirmative action manager does not relieve officers of their duty to be aware of and comply with the intent of equal opportunity legislation. Courts have held that discrimination at any level is not acceptable. And affirmative action applies to all aspects of the job, not just hiring and termination. The administrative officer must monitor daily activity and root out any discriminatory practices.

Furthermore, the courts have held that actions resulting in unequal treatment, even though not blatantly discriminatory or intentional, are as unacceptable as any overt discrimination. This implies that organizations are judged not by the process they choose to ensure nondiscriminatory hiring practices, but by the end results of those practices (see *Uniform Guidelines on Employment Selection Procedures,* FR#2809C, August 1978).

EQUAL EMPLOYMENT OPPORTUNITY AND AFFIRMATIVE ACTION

Officers often ask about the difference between equal employment opportunity and affirmative action. Offices of equal employment opportunity are regulatory/advisory agencies at the state and federal levels that monitor employment practices and investigate discrimination complaints. Affirmative action, on the other hand, is a voluntary action on the part of the employer to ensure that hiring practices meet the intent of equal employment opportunity acts and executive orders. The intent of equal opportunity is to correct the effects of past discrimination, to eliminate present discrimination, and to prevent future discrimination in hiring and employment (Figure 2.2).

Figure 2.2 Under equal employment opportunity, all types of people have the right to be considered for a job.

Affirmative action programs are needed by employers to avoid legal difficulties under equal employment opportunity acts. An affirmative action program is not required by law, but its absence can be construed as prima facie evidence of discriminatory hiring practices. Therefore, most organizations have some form of affirmative action program. The basic elements of an affirmative action program are a management commitment to affirmative action and an affirmative action plan. The management commitment is usually demonstrated by a clear policy statement, but the affirmative action program is somewhat more involved. The program should include specific, documented procedures that define how any past discriminatory effects will be corrected, how present discriminatory practices will be ended, and how future discriminatory practices will be prevented.

Affirmative action often seems overwhelming. Some ask, "What if we decide to play the odds?" There is no requirement to have a program or to pursue affirmative action. The only real requirement is that the employment practices of an organization be free from discrimination. A hypothetical example follows:

In City X, the city management felt that the current practices of employment were not discriminatory now or in the past. Therefore, City X opted not to develop an affirmative action program. Administrators continued their regular employment practices, there were no complaints from employees or prospective employees about the employment practices, and everything was fine. But consider the case of City Y: City Y also opted not to develop an affirmative action program. But an applicant turned down for a position filed a discrimination complaint — and won. The applicant got the job, back pay, and seniority because City Y could not document its nondiscriminatory practices.

The point: the courts have considered the lack of an affirmative action program as evidence of discrimination. An affirmative action program can help a department avoid costly litigation.

AFFIRMATIVE ACTION PROGRAMS

An affirmative action program is composed of four basic parts:

- Management commitment to affirmative action
- A written affirmative action plan
- Nondiscriminatory hiring, promotion, and termination practices
- Affirmative action documentation

Management Commitment

If the governing entity is not totally committed to an affirmative action program, it is likely it will be involved in a discrimination lawsuit — and lose. It is not economically feasible to ignore affirmative action. The organization eventually will be required to pay for legal proceedings, correct the effects of past discrimination, and comply with court orders.

Every organization has the opportunity to set affirmative action goals and objectives, and when they are pursued with reasonable earnestness the organization is not likely to be faced with adverse legal requirements. It will also be in the best position to employ and promote the best candidates for vacancies. But when a city fails to develop or pursue affirmative action goals, it may be subject to court-mandated quotas, which generally supersede the goals or guidelines an organization may have imposed on itself and require that hiring and promotional practices be aimed at specific sectors of the community (Figure 2.3). The fire department could still employ the best applicant but would not be allowed to consider persons who did not meet quota requirements. It is obviously in the department's best interest to incorporate affirmative action goals and objectives in departmental recruitment.

Race and Sex Discrimination

FEDERAL APPEALS COURT UPHOLDS BUFFALO POLICE AND FIRE HIRING GOALS.

Interim goals for correcting racial and sexual imbalance in the Buffalo fire and police departments were established by the District Court. Out of every four applicants for either department, two must be black and one must be female (a black female counts twice).

The Second Circuit affirmed the hiring goal. As to long term goals, no action was taken. The city maintained they would be unnecessary upon adoption of a validated entry test. The issue was premature, said the court, because no such test has proven successful. **United States v. City of Buffalo,** 24 FEP Cases 313, 633 F. 2d 643 (2nd Cir. 1980). FP# 3692 D.

Figure 2.3 Well-developed affirmative action plans are good preventive measures against legal actions such as court-ordered quotas. *Courtesy of Fire and Police Personnel Reporter, October 1981.*

Goals are best met by aggressively recruiting to find the best job candidate available and making a special effort to recruit qualified minorities and disabled persons. Hiring quotas suspend regular hiring practices and limit job applicants to groups that are underrepresented in an organization. The best medicine to prevent court-ordered quotas is a management commitment to affirmative action.

THE AFFIRMATIVE ACTION PLAN

The first step in drawing up an affirmative action plan is to determine where the organization stands in respect to the avail-

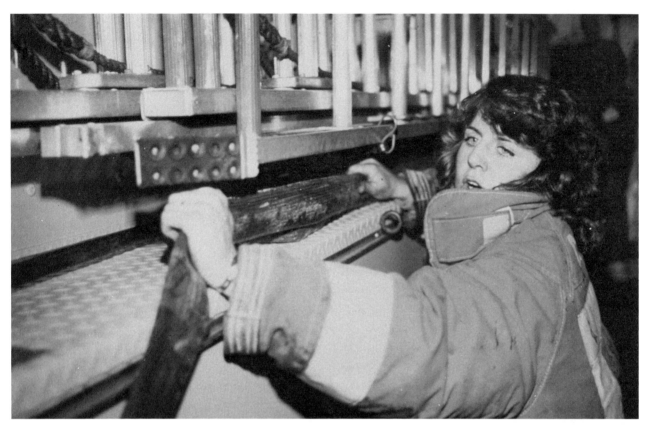

Figure 2.4 The goal of affirmative action programs is to have a work force that reflects at every level the sexual and ethnic composition of the surrounding area. *Courtesy of Larry Phillips and Denise Demars.*

able work force. Theoretically, the composition of the department work force should match the composition of the available work force of the surrounding community (Figure 2.4). A good place to begin is with the state employment service to determine the proportions of minority and disabled persons in the jurisdiction. In most instances, this study will be conducted on a community-wide basis and the affirmative action program organized on that level. However, this does not release fire department officials from the duty to check their own department employment practices to determine if they are following the affirmative action guidelines.

Community percentages should be compared with a composite of department personnel. If the percentages are not relatively equal, it can be assumed that the disparity is the result of blatant or inadvertent employment discrimination. Equalizing these percentages will correct the effects of any past discrimination.

Even if the percentages are relatively equal, it does not mean that equal employment opportunity has been achieved. Administrators must then examine the dispersion of minority personnel throughout the organization. They should be distributed uniformly. There should be a proportional amount of minority personnel in management and line positions. Paying 25 female clerk-typists minimum wage will not balance 25 male battalion chiefs on executive salaries.

After examining these two basic issues, administrators can formulate a plan to correct the effects of any past discrimination and prevent any future discrimination.

Affirmative Action Implementation

There are six basic steps to implement an affirmative action program:

- Demonstrate management support by issuing a written equal employment policy and affirmative action commitment.

- Make a top official responsible for implementing and maintaining the program.

- Publicize the commitment and policy.

- Monitor employee proportions by job classification and department.

- Pursue goals and timetables to improve the hiring and promotion of minority employees.

- Develop and implement specific programs to achieve and maintain goals.

Because of today's affirmative action climate, each affirmative action program will have active and reactive facets. The active facet attempts to ensure compliance with the intent of equal employment mandates, administers the program and tries to reduce the impact of possible legal actions against the department. The reactive part of the plan develops documentary evidence that hiring, promotion, and termination have been conducted legally.

There is no "correct" way to develop an affirmative action program. The test of the program is its result, not its organization or format. Therefore, most attention will focus on the reactive or documentary parts of the program. Keep in mind that if the city is sued, it must prove that it was not discriminating. Unfortunately, it is practically a case of being guilty until proven innocent, and the way a city can best prove its innocence is with proper documentation.

Affirmative Action Documentation

Documenting affirmative action begins by developing job descriptions. Hiring criteria should be based on people's ability to do the job; therefore, the department must know exactly what job it wants done. Like developing an affirmative action program, there is no "right" way to write job descriptions.

Some jurisdictions, such as Oklahoma City, use general job descriptions. There, the department's job description for an entry-level firefighter requires that the applicants be in good physical shape and able to learn, follow orders, communicate orally and in

42 CHIEF OFFICER

FIREFIGHTER RECRUIT

BOTH MEN AND WOMEN ARE ENCOURAGED TO APPLY

SALARY
$16,786 - $17,722 annually, paid in biweekly payments - 56 hours per week

DUTIES
Firefighter Recruit is a trainee class for protective service work in the Fire Department. Recruits are sent to the Phoenix Fire Academy for three months for training in firefighting operations, emergency medical treatment, fire code enforcement, fire prevention practices, public relations, physical fitness, and other related subjects. After successful completion of training at the Fire Academy, recruits are _____ of Firefighter. The annual salary range at this time for Firefighters is $20,363 _____ Phoenix Fire Department. Firefighters may be re_____

EXPERIENCE AND _____
Must be at least 18 y_____
two documents, suc_____
hire.) Completion of _____
training at the Fire _____
physical and menta_____
Retirement System_____
correctable to 20/_____

RESIDENCE
Must live within _____
map showing thi_____

EVALUATION _____
A valid I.D. card _____
gain admittanc_____
and at the time _____

Written 85% _____
Physical Agili_____

WHEN TO A_____
January 9, 1_____
or after thes_____
300 W. Wa_____

REFEREN_____
6101.9 - F_____
SC - PD #_____

EXAMPL_____
• At the s_____ and cli_____
• Perfor_____
• Admin_____ pulmonary fun_____
• Participates in fire drills and attends _____
• Performs general maintenance work in the upkeep of the _____
• Inspects building and premises for compliance with fire laws;
• Checks on complaints and aids in the investigation of arson cases;
• Inspects commercial and noncommercial buildings;
• Signs citations and gives testimony in court in connection with fire code enforcement activities;
• Organizes fire brigades and conducts fire drills in institutions and commercial establishments;
• Attends public gatherings to ensure observance of fire safety requirements;
• Participates in station radio tests;
• Conducts tours of station houses for scout, civic and other interested groups;
• Operates Computer Aided Dispatch equipment.

FIREFIGHTER RECRUIT

APPLICATIONS MAY BE ACCEPTED AT FIRE DEPARTMENT HEADQUARTERS
428 West California
Fire Chief's Office

JOB SUMMARY: This job will be located at one of 31 existing Fire Stations, under the general supervision of a Fire Captain. Firefighters will perform individually and/or as a member of a fire fighting team and will participate in fire suppression activities including fire fighting, rescue, first aid, ventilation, forcible entry, salvage, and drill ground training sessions on a regular basis to continuously update their skills and knowledge required for effective work performance. Decisions are made with guidance of a supervisor and on an individual level and have a direct effect on life and property. General job functions include (1) fire fighting, (2) first aid assistance, (3) continuous training.

JOB REQUIREMENTS:
1. Good physical condition.
2. Good eye condition.
3. Ability to learn.
4. Ability to work safely in an abnormal atmosphere.
5. Ability to learn the use of both manual and mechanical tools.
6. Ability to communicate both orally and in writing.
7. Ability to work as a part of a team and alone.
8. Ability to adapt to various situations.
9. Ability to work in adverse weather conditions.
10. Ability to work irregular hours.
11. Willingness to conform to Uniform Appearance Code.

Figure 2.5 Some job descriptions are general; others are specific with regard to duties and qualifications. Evaluations are based on the results of written and physical exams. *Courtesy of City of Phoenix and Oklahoma City Fire Departments.*

writing, work safely as part of a team at irregular hours in all kinds of weather, and adapt to different situations. Other departments prefer more specific job descriptions (Figure 2.5).

Both types of descriptions have been validated in court. The important legal aspect is not the type of description, but how the department tested for the position. In other words, how valid was the test? Test validity has become tremendously important in recent discrimination cases.

Test Validity

Test validity means that the testing procedure clearly demonstrates the applicant's ability to perform a job-related task (Figure 2.6). There are three basic types of entry-level testing: written, oral, and physical. The degree of liability the department faces depends on the content and application of the tests.

Figure 2.6 Tests given to applicants must measure skills needed for performing a job. Here, an applicant is demonstrating strength and agility by hanging a smoke ejector.

Another important testing requirement is that the test cannot require the applicant to perform at a higher level than any person currently holding the position. Before giving a test, the department should be sure that every person holding the position can pass the test. Some have tried unsuccessfully to convince the courts that their departments are trying to upgrade the level of fire service by increasing entry standards. The courts have said that service levels must be improved within the ranks before increased ability becomes an employment criterion.

The final requirement is an analysis of the test's impact on the employment of minority and women applicants. If the test reduces the proportions of such applicants hired, the department might be sued. Then the department will have to prove the test was not discriminatory. However, the courts have upheld the validity of tests with an adverse effect on minority employment if the tests are clearly job related (see Bridgeport Guardians vs. Bridgeport Police Department, 431 F. Supp. 931: D.Conn. 1977, Jan. 13).

There is no standard way to prove job relatedness. However, there are basic criteria that can help:

- Does the person holding the position perform the task regularly? Or, if the position is vacant, did the previous occupant?
- Does the test in fact measure the individual's ability to perform that task?

There have been studies to determine if a task is job related. The most noted are "A Study of the Fireman's Occupation" by the University of California in 1968 and the Michigan Municipal League's "Firefighter Selection Research Project." Researchers asked firefighters to fill out extensive questionnaires about their job, rating specific fire fighting tasks by the number of times they performed the task and by the task's relative importance to the overall job. The studies determined which jobs are done by most firefighters in the course of their daily work. Departments can then infer that an applicant who demonstrates a higher ability to perform that type of task would be a better candidate for a position than an applicant with lower ability. Obviously, departments cannot test for every fire fighting task. Therefore, a study may be necessary to determine exactly what skills are needed to perform the most important or frequently required tasks. In many cases, these tasks will require similar abilities or skills. The department may need an industrial engineer or psychologist to test some of these abilities, but others, such as ladder carrying, can be tested directly.

A secondary issue: if the task is critical and is performed often, must applicants possess the skills when hired, or can the skills be learned on the job? This question has forced many fire departments to reassess their testing criteria to test only for abilities required to develop the needed skills. Is there an academy training period, or will applicants receive on-the-job training? Must applicants perform the tasks before formal training?

After departments select the abilities to use as employment criteria, officials must carefully examine testing methods. Here's the question the courts might ask: "Did the test really measure the individual's ability to perform that task?" The legal issue can be complex, and judges make the final decisions.

Court trends indicate that if departments can show that their entrance tests are job related and that the tests measure the skills required for the job, then the tests are valid and their effect on minority hiring cannot be a basis for discrimination suits. However, there are still some criteria that must be met before the department can avoid discrimination complaints. The courts have shown a willingness to accept validated test results only when the department can show that a proportional amount of minorities took the test. If the department was discriminatory in its recruitment practices, the validity of the test is no longer the issue.

Recruitment

The legal question concerns the positive tangible steps the organization is taking to ensure equal employment opportunity. This means that the department is obligated to ensure that adequate numbers of minority applicants have the chance to take the test. Furthermore, fire departments must actively recruit minority applicants. No longer is it acceptable to post the position on the city hall bulletin board and ask firefighters to tell their friends and relatives about the openings. The fire department must go into the community and seek out minority candidates. The logic behind this is simple. Minority applicants might feel that fire department employment is not open to them and assume the application procedure is the department's token act to meet equal employment opportunity requirements. Convincing minorities that this is false and that they have an equal chance to be employed is the department's job.

Ideally, fire department tests would be taken by a proportional cross section of the community work force, and the test would identify the best candidates who would also represent a cross section. Then the department would probably never be involved in a legitimate court case. But keep in mind that nothing can keep a department from being taken to court. There will always be a possibility that applicants will feel they have been discriminated against and file suit. This does not mean they will win. The plaintiff's success will depend more on the department's documentation of its employment system than any other factor.

One of the recruitment tasks involved in equal opportunity is documenting minority involvement.

This can be touchy, because some applicants will worry about how the department will use information from job interviews. For example, is it discriminatory to ask an applicant if he or she is a minority group member?

The answer can be yes or no, depending on why the question was asked and what the information will be used for. If the answer is a criterion for employment, the question is discriminatory. If the question was asked only to document the department's efforts to recruit minorities, the question is acceptable. Because departments must document their efforts to recruit equitably, they must obtain this information from the applicants (Figure 2.7).

Employment Criteria

One of the better ways to obtain the documentation is to distribute an affirmative action questionnaire at some point in the application procedure. The affirmative action officer should explain the questionnaire, emphasizing that the department will not use the information for employment purposes. If possible, the form should be a "checkoff" type that doesn't require a signature. The forms should be numbered for control, but handed out randomly. This will help document the fact that everyone who applied filled out a form and that the department did not bias the sample. In addition, the randomness demonstrates that the information could not have been used to make a hiring decision.

The department cannot use the same random technique for the handicapped and Vietnam-era veterans because these persons may be granted employment preference in some cases. Information on their status must be retained with their job applications.

Once the department gets the information in a way that minimizes its improper use, administrators should tabulate it. They should catalog the forms to indicate the percentage of each population group that filled out the questionnaire. Administrators can then compare these statistics with community statistics to ascertain the success of the department's recruiting efforts. The statistics do not have to match exactly, but officials should suspend the testing procedure if there is a wide variance, at least until an appropriate number of applicants can be recruited from underrepresented groups.

CIVILIAN EMPLOYEES IN THE FIRE DEPARTMENT

Recent court decisions have also affected the hiring of civilian employees. Traditionally, almost everyone the department hired had to pass the firefighter's entrance exam. Many courts

PRE-EMPLOYMENT INQUIRY GUIDE

SUBJECT	LAWFUL PRE-EMPLOYMENT INQUIRIES	UNLAWFUL PRE-EMPLOYMENT INQUIRIES
NAME:	Applicants full name.	Original name of an applicant whose name has been changed by court order or otherwise.
	Have you ever worked for this company under a different name?	Applicant's maiden name.
	Is any additional information relative to a different name necessary to check work record? If yes, explain.	
AGE:	*Are you 18 years old or older?	How old are you? What is your date of birth?
RACE OR COLOR:		Complexion or color of skin.
HEIGHT:		Inquiry regarding applicant's height.
WEIGHT:		Inquiry regarding applicant's weight.
MARITAL STATUS:		Requirement that an applicant provide any information regarding marital status or children. Are you single or married? Do you have any children? Is your spouse employed? What is your spouse's name?
SEX:		Mr., Miss or Mrs. or an inquiry regarding sex. Inquiry as to the ability to reproduce or advocacy of any form of birth control.
HEALTH:	Do you have any impairments, physical, mental, or medical which would interfere with your ability to do the job for which you have applied?	
	Inquiry into contagious or communicable diseases which may endanger others. If there are any positions for which you should not be considered or job duties you cannot perform because of a physical or mental handicap, please explain.	Requirement that women be given pelvic examinations.
CITIZENSHIP:	Are you a citizen of the United States?	Of what country are you a citizen?
	If not a citizen of the United States, does applicant intend to become a citizen of the United States?	Whether an applicant is naturalized or a native-born citizen; the date when the applicant acquired citizenship.
	If you are not a United States citizen, have you the legal right to remain permanently in the United States? Do you intend to remain permanently in the United States?	Requirement that an applicant produce naturalization papers or first papers.
		Whether applicant's parents or spouse are naturalized or native born citizens of the United States; the date when such parent or spouse acquired citizenship.
NATIONAL ORIGIN:	Inquiry into languages applicant speaks and writes fluently.	Inquiry into applicant's (a) lineage; (b) ancestry; (c) national origin; (d) descent; (e) parentage, or nationality.
		Nationality of applicant's parents or spouse.
		What is your mother tongue?
		Inquiry into how applicant acquired ability to read, write or speak a foreign language.
EDUCATION:	Inquiry into the academic vocational or professional education of an applicant and the public and private schools attended.	
EXPERIENCE:	Inquiry into work experience.	
	Inquiry into countries applicant has visited.	
ARRESTS:	Have you ever been convicted of a crime? If so, when, where and nature of offense?	Inquiry regarding arrests.
	Are there any felony charges pending against you?	

*This question may be asked only for the purpose of determining whether applicants are of legal age for employment.

Figure 2.7 Although departments need to document the types of candidates who apply for positions, certain questions are irrelevant to the job and may not be asked. *Courtesy of Michigan Department of Civil Rights.*

have invalidated this practice, pointing out that many jobs in the fire department can be performed by a person without fire fighting experience and that it is unreasonable to require applicants to have fire fighting skills. Such jobs include dispatch, fire prevention, public relations, and public education (Figure 2.8).

In fact, many fire departments are now going outside the ranks to bring in the expertise needed for special programs. When the department instituted a public school education program, for example, officials using the traditional approach sought a firefighter with a compatible interest. If they could not find one, they "volunteered" one. Fire department administrators are now approaching the problem differently by seeking employees with expertise from outside the fire department and putting them on the staff in civilian positions. This approach has advantages. Civilians normally handle these tasks better because their expertise is the criterion on which they are hired, not their fire fighting experience. This approach also brings new skills and perspectives into the fire department, allowing officials to utilize the skills of people who might otherwise have been excluded from fire department employment. There is also an economic advantage in states where worker compensation premiums for uniformed employees are higher than for civilian employees. But administrative officers must make sure that any civilians they hire understand the specialized concerns of the fire department.

Figure 2.8 Fire departments have made excellent use of civilians with special expertise in such areas as public relations, fire prevention, and public education.

Fire Company Staffing

3

CHAPTER 3 OBJECTIVES

1. Identify and document local factors affecting fire department staffing levels.

2. Identify alternative methods for maintaining adequate first response capabilities.

3. Analyze training procedures for maintaining maximum efficiency of companies and make recommendations for improvement.

Chapter 3
Fire Company Staffing

Distributing and staffing fire companies are often decided arbitrarily in the fire service. There is some scientific data available on which to base decisions, but current opinion depends largely upon traditional coverages accepted by fire insurance underwriters and dignified by promulgation as standards.

HISTORICAL STAFFING TRENDS

When firefighters pulled apparatus to the fire and pumped water by hand — one of the most strenuous forms of exercise — each company had to have a large number of members to meet the demanding work load (Figure 3.1). The annual report of the New York Fire Department for 1862 shows that engine companies had from 30 to 40 members, hook and ladder companies had slightly fewer, and hose companies had 20 to 30. All were volunteers.

Figure 3.1 After steam engines came into use, the number of firefighters dropped sharply.

Three years later, when the paid force supplemented the volunteers and horse-drawn steam engines were used, the number of company members dropped greatly. For example, Engine Company 24 of the volunteer department had 37 members, but when the same company became a paid unit the roster went down to 12. There were eight firefighters, an engineer, a driver, and two officers. How the number of firefighters was determined is no longer clear, but it became the basis for deciding the on-duty strength of fire companies in major cities for the next 80 years.

Until 1919, paid firefighters worked a system known as "continuous duty." They worked 24 hours every day, except for three hours a day for meals and a day off every 10 days. With this system there were usually seven firefighters, an officer, an engineer, and a driver on duty.

This was the situation in major American cities in 1914 when the National Board of Fire Underwriters made the study that became the basis for the first Grading Schedule, so the board naturally proposed similar company strength. Then by 1915 when the Grading Schedule was promulgated, gasoline-powered apparatus were replacing horse-drawn vehicles, a transition that would be almost complete by 1925. Also, water systems and hydrant spacing had been improved greatly since 1865. Still tradition died hard. The seven-member company remained the standard that fire departments tried to achieve until the Municipal Grading Schedule changed in 1974.

Each of America's major wars ended with an almost immediate reduction in fire department working hours. Right after World War I administrators in most American cities replaced the continuous duty system with the two-platoon system, cutting the work week to 84 hours. To maintain the same on-duty strength, administrators had to double total company membership. As administrators replaced steam engines, they also phased out the engineers because the driver could also operate the pump. So between the two world wars, fire departments tried to staff companies with an officer, a driver, and seven firefighters on each of two platoons. This required an 18-member company.

World War II's forced reduction in personnel convinced many city officials they could meet their needs with smaller companies. Consequently, when the end of the war was followed by a shortened work week, most cities abandoned the effort to maintain traditional company strength. In some cases the reduction was drastic. Three- and four-member companies became more common than six-member units, and it appeared that the seven-person company was gone forever.

The Grading Schedule was completely revised in 1980 as the Fire Suppression Rating Schedule. The new schedule avoids mentioning specific numbers for minimum company strength, but offers a formula which gives a maximum credit for six members on engine and ladder companies.[1] The formula is as follows:

$$CCP = \frac{2\frac{1}{2}(OM + VM/3)}{EE + EL + 0.5(ES) - SC}$$

Where CCP = Credit for company personnel
       OM = On-duty strength
       VM = Volunteer or call members responding on first alarms
       EE = Existing engine companies
       EL = Existing ladder companies
       ES = Existing service companies
       SC = Surplus companies

FIRE COMPANY STAFFING STUDIES

What is the ideal number? Studies analyzing how the inputs and outputs have changed over the years and how the former affect the latter are rare in current literature. The influence of the Grading Schedule has faded rapidly in recent years, replaced in some places by labor contracts requiring minimum staffing. Governing bodies are also trying to hold down rapidly increasing personnel costs.

Two reports of attempts to base needs on actual tests involve a set of experiments by W. E. Clark in Wisconsin in the early 1960s[2] and another by the Dallas Fire Department in 1969.[3]

Wisconsin Study

Clark emphasized that the Wisconsin tests measured time only. He mentioned but did not measure other factors such as fatigue, and the tests all dealt with one type of hypothetical fire. From two to six firefighters tried to accomplish the same objectives. The completion times varied immensely with the staffing but the differences were not exactly proportional, showing that increasing the number of people quickly reaches the point of diminishing returns.

Clark also found that a change in technique can be as influential as a change in staffing.

Dallas Study

Dallas patterned its tests on Clark's but introduced several new variables. There were more tests of differing complexity, and Dallas compared times when different companies tried each task. Critics complained that Dallas tried to prove the need for larger companies, but testers reported the company times candidly. Tests of different fireground operations were conducted using three, four, five, and six firefighters. The averaged percentages show efficiency increasing proportionally with staffing. It was determined that five firefighters were the minimum necessary to perform most operations efficiently and without undue fatigue.

Staffing Study Conclusions

The results indicate that quality can make a difference. A company composed of a team of physically fit, well-trained firefighters will perform better than a group less carefully selected and not as well trained.

Changes in technique can also make a difference. One technique Clark described, which sharply reduced time and effort, was using 1½-inch (38 mm) hose instead of 2½-inch (65 mm) hose to reach a fourth-floor fire. Not surprising, but this was the first time someone measured such a change and used it to decide company strength. How would other technological changes affect staffing? Some, such as lighter hose, portable ladders, and automated equipment, might reduce physical effort. Tactical problems could be reduced by improving building construction, building in protective features, and prohibiting conditions that force fire departments to identify some structures as target hazards. On the other hand, technological changes that increased the potential for fires or increased a fire's severity would demand more staffing.

FACTORS AFFECTING STAFFING

Department administrators must realize that changing factors influence company size, that they are not static. Chief among these are budget considerations, tactical policies, crew safety, and labor agreements. Other duties performed by fire fighting personnel may also play a part in company size: EMS, hazardous material spills, rescue operations, and fire prevention.

In many departments, company officers participate in suppression tasks if staffing is low. The major concern in this situation is whether lack of supervision lessens the effectiveness of fireground operations.

Company Strength

Since the tasks confronting a fire company vary from district to district, administrators should adjust company strength accordingly. Does a company covering an area of well-spaced single-family dwellings need the same staffing as one covering a heavily industrialized district? Logically, administrators should adjust company strength to the tasks the unit will ordinarily face (Figure 3.2). It certainly takes more effort to stretch 2½-inch (65 mm) hose to the fifth or sixth floor of New York's tenements and lofts than to operate a 1½-inch (38 mm) line in the one-story dwellings that cover most of Phoenix. The same is true of truck company duties.

Figure 3.2 Personnel requirements will vary depending on the type of area to be protected.

Most large cities have more firefighters assigned to a ladder company than to an engine company. The reasons could lie in the types of construction more commonly found in large cities and the more strenuous tasks ladder companies are assigned. In many smaller cities the reverse is the norm.

The first-alarm assignment also varies. In one study of 18 large cities, the lowest reported assignment consisted of two engines and one truck, and the highest was four engines and two trucks. The averages were 3.27 engines and 1.77 trucks in high-value areas and 2.77 engines and 1.27 trucks elsewhere.[4] Many cities during the past decade have increased their first response assignments to compensate for reduced crew sizes. The 1980 ISO Rating Schedule acknowledges that first-alarm requirements vary with the fire and therefore derives its recommendations from formulas based on "fire flow," the amount of gallons per minute insurance engineers decide firefighters need to put out a fire[5] (Table 3.1 on next page).

Initial response may also vary with weather conditions and type of alarm. Since they are often false, many departments respond with only one engine to alarms from street boxes. Other types of mechanical alarms with a history of false alarms may also receive a smaller initial response. Severe weather conditions

**TABLE 3.1
Basic Fire Flow**

GPM	LPM	Number of Needed Engine Companies
500-1000	1900-3800	1
1250-2500	4750-9500	2
3000-3500	11,400-13,300	3

Source: ISO Fire Suppression Rating Schedule

have caused departments to increase on-duty staffing or to respond with additional units.

Staffing on companies has dropped from an average of four or five members during the 1970s to three or four members during the eighties. Smaller cities often respond with even fewer personnel. A study by Centaur Associates, Inc. for the U.S. Fire Administration examined 171 cities of more than 100,000 population. The study showed that 55 percent of the departments responded with four persons on an engine company, 35 percent with three, and 5 percent with two.[6]

Many fire departments get along with lightly-staffed first-alarm assignments because of budget constraints, but quite a number of fire officers feel that larger forces are unnecessary. In one study, 74 percent of those queried felt that 12 firefighters were enough to combat a well-advanced fire in a seven-room house.[7]

Out of the same group a slight majority thought that not more than 18 would be needed for a basement fire under a grocery store in a two-story building of ordinary brick, 20 feet (6 m) x 90 feet (27 m). For a fire in a two-story factory of ordinary brick, 80 feet (24 m) x 100 feet (30 m), 48 percent thought that no more than 23 firefighters would be enough, while 24 percent felt that 12 could handle it.

Of course, there was a wide range of opinion, and some people differed from others by substantial margins. The questions were asked of three groups in widely separated sections of the country: New England, the Deep South, and the Midwest. The variety in the answers seemed to be based more on size of department than on regional differences. Those from smaller departments felt less staffing was needed.

Response Distance

On the matter of response distances, the survey respondents also had differences of opinion as to what was acceptable. Sixty-six percent said the first-due engine dispatched to a fire in a high-value district should travel no farther than one (1.61 km) to three miles (4.8 km); 30 percent said less than one mile (1.61 km). For residential districts, only 8 percent said response distance should be less than one mile (1.61 km), 71 percent said one to three, and 11 percent said four. Another study of 1,400 fire departments showed that the response time for fully-paid departments averaged 3.8 minutes. The average time for the longest run to areas with structures was 7.9 minutes.[8]

What is the conversion factor to translate miles into minutes? Some studies report a fairly consistent pattern of two minutes per mile. This can vary considerably with time of day and street patterns. Another study of several cities shows variance from city to city, with the average being about three minutes per mile.[9] Some firms market computer software that analyzes factors fire departments should use to locate stations.

The ISO Rating Schedule says built-up areas of a city should have a first-due engine within 1½ miles (2.41 km) and a ladder company within 2½ miles (4 km).

Work Week and Leave Allowances

Two other factors have to be considered with response distances and company strength to determine total staffing needs: the work week and annual vacation and sick leave.

To maintain four persons on each of three platoons on a 56-hour work week requires additional personnel to cover sick leave and vacations, usually about 10 percent of the company.

A 42-hour week for four people requires one more person per company than the 56-hour week before the 10 percent is added.

The National Fire Protection Association, after reviewing fire department records, claims that a 20 percent additional factor is more realistic, and that 3.75 firefighters are needed to keep one firefighter per shift on duty for a 56-hour week, five per shift for a 42-hour week.[10]

The same source suggests that cities usually need 2.5 firefighters to keep .5 firefighter on duty per 1,000 population. In 1974 twenty-five fire departments on a 42-hour week averaged three firefighters per 1,000 population.

In 1981 residents of a Texas city voted by a slight majority that the city should employ at least 1.6 firefighters per 1,000 population on a 56-hour week, somewhat under the apparent national average of 1.68.

Averages when used with other reliable data work as benchmarks and verifiers. But averages can be skewed by extremes, and they are not always applicable. The more reasonable approach to decision making bases requirements on specific needs determined by analysis. Use averages and other national data for comparison to discover extreme deviations, which should then be re-examined.

ALTERNATIVE SOURCES OF COMPANY STAFFING

Personnel requirements should be based on the total strength of the first-alarm assignment, not individual company strength. Once this has been determined, consideration can then be given to volunteers and automatic mutual aid. The number of firefighters arriving in a given time is more important than where they came from or how they got there.

Some fire departments have personnel squads that respond over larger areas than engine or truck companies to build up total first-alarm strength. Other cities have used police officers or "public safety" officers (a combination police officer and firefighter) to drive police cars to fires and fight fires when needed. Neither concept, especially the latter, seems to have lasted long. Combining police and firefighters has been tried and abandoned in more than 100 cities.

Some smaller cities have used employees from other city departments such as public works to supplement regular firefighters. Not all chief fire officers are satisfied with this arrangement, but others think it is necessary when volunteers are away from town during the day. Some towns have granted high school students standing permission to leave school for fire alarms. Other departments have hired paid firefighters to work when volunteers are away.

One rather obvious solution disregards political boundaries and assigns companies from other jurisdictions to respond jointly under automatic mutual aid. This has worked in the Washington, D.C., and Boston areas. Several areas have overcome the barrier of legal liability, and the concept has proven economically and strategically feasible.

In the 1970s the phenomenal growth of emergency medical service within the fire service created unprecedented problems in distributing personnel. A vivid example is a midwestern city of about 45,000 that housed three engine companies and a truck company in three stations. Three people were on duty in each company. When the department put an ambulance in each station, two firefighters from the engine company went on each ambulance run, leaving one behind to run the engine. If a fire occurred when two ambulances were out, only eight firefighters protected the whole city.

On the other hand, some fire chiefs have obtained substantial increases in personnel to provide emergency medical service without diminishing their fire fighting strength.

In some fire departments, emergency medical technicians respond to alarms. They don protective clothing and fight fires in some cities. In others, they perform standby or support duty that leaves them clean and free for medical calls (Figure 3.3).

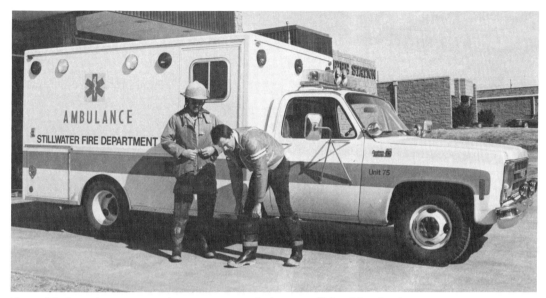

Figure 3.3 Some departments assign emergency medical personnel "clean" duty; in other departments, they are assigned full suppression duties.

Whatever the situation, fire departments know that company staffing should be somewhere between one and six, the upper limit for credit under the suppression schedule. They know the city can have one fire station or, if it has the money, a station on every corner. Departments have to find their "right" answers as scientifically as possible, relating their resource requirements to legitimate needs. Chief officers have no tried and proven scientific formulas to apply, but there are several factors they can use to reach and justify their decisions. Below is a partial list.

- The level of service and response times desired
- The amount of emergency medical service provided
- The number of support programs wanted
- The quality and type of apparatus and equipment
- The level of personnel safety required
- The quality of training programs
- The fire-load profile of the community

Unfortunately, city budgets have more effect than need when it comes to staffing. Fire department administrators with a reasoned, convincing approach will fare better in competing for the tax dollar. Those who have to do with less must allocate personnel according to need.

References for this chapter may be found in Appendix B.

Labor Relations

4

NFPA STANDARD 1021
STANDARD FOR FIRE OFFICER
PROFESSIONAL QUALIFICATIONS

Fire Officer VI

7-3 Labor Relations.

7-3.1 The Fire Officer VI, given a summary of the types of grievances which may confront the officer, shall:
    (a) describe the methods by which the officer can detect and analyze the cause of grievances
    (b) describe the method by which the officer establishes an equitable grievance procedure
    (c) describe the methods by which the officer adjusts and handles grievances.*

7-3.2 The Fire Officer VI, given a summary of the laws, ordinances and procedures established by the jurisdiction to govern employee selection and promotion, wages, benefits, and other conditions of employment, shall demonstrate knowledge of how to administer a labor relations program.

The above NFPA standards are addressed from a general management perspective.

*Reprinted with permission from NFPA No. 1021, *Standard for Fire Officer Professional Qualifications*. Copyright 1983, National Fire Protection Association, Boston, MA.

Chapter 4
Labor Relations

There was a time when chief officers did not have to be concerned with employee organizations, collective bargaining, strikes, and other job actions. That time is long past. The trend today is toward unionization. The International Association of Fire Fighters, which claims to represent about 95 percent of the organized firefighters in the United States and Canada, says its membership represents 80 percent of all full-time career firefighters. The IAFF says the remaining 5 percent belong to local independent unions or locals affiliated with the Teamsters or other unions. The association notes that a number of volunteer firefighters belong to unions representing workers in industries employing the volunteers. The National Fire Protection Association points out that even unorganized departments have implied labor contracts based on verbal agreements between chiefs, supervisors, and firefighters; traditional practices; state laws and local ordinances; and handbooks for supervisors and firefighters. So a knowledge of labor relations is now vital for chief officers.

DEVELOPMENT OF PUBLIC EMPLOYEE UNIONS

The labor union phenomenon among public employees is of relatively recent vintage. The laws granting employees the right to organize unions and to bargain collectively were passed as part of President Franklin Roosevelt's response to the Great Depression. But neither the National Labor Relations Act nor any of its major amendments covered government employees at any level. The National Labor Relations Board, which the federal legislation created, certifies labor unions as collective bargaining agents and tries to make sure that management and labor operate fairly. Public employees are specifically excluded from NLRB jurisdiction. Congress apparently felt that public employees were not involved in commerce and competition and therefore had no need for unions. Thus, some states, beginning with Wisconsin in 1959, passed their own laws to regulate the organization of public employees. Most of the laws, which a number of states have today, allowed some if not all public employees to organize and bargain collectively, but denied them the right to strike (Figure 4.1).

Figure 4.1 Although firefighters have the right to join unions, it is illegal for them to go on strike.

A number of factors encouraged the development of public employee labor unions. One was the civil rights movement of the 1950s and 1960s as blacks and other minorities began to assert their constitutional rights. The development at the national level was spurred by executive orders from Presidents Kennedy and Nixon. Kennedy in 1962 set up rules under which federal employees could bargain collectively. The order also listed the rights of management, outlined how to handle grievances, and spelled out how unions were to be recognized. Nixon's order set up the Federal Labor Relations Council that now does for federal employees what the NLRB does for private sector employees.

Organization progressed during the 1960s and 1970s, sometimes over the strong opposition of administrators. Even though the right of public employees to organize became accepted, some in and out of the judiciary felt that the right should not extend to public safety employees because the taxpayers and government needed to be sure of their loyalty. But as Charles S. Rhyne points out in his book, *Police and Firefighters: The Law of Municipal Personnel Regulations*, the courts eventually rejected that argument and ruled that firefighters and other public safety employees have a constitutional right to organize and join labor unions. Judges, noting that the fire service is a quasi-military operation, have upheld legislative bans on strikes. The courts have added that the quasi-military nature of the fire service, which demands discipline and loyalty, is not reason enough to prohibit unionization.

However, courts have occasionally qualified the rights of firefighters to join unions. Some cities have beaten back challenges to rules limiting membership in groups advocating the right to strike despite legal bans. In some cases, the courts have upheld the right of cities to prohibit public safety officers from joining unions that included other categories of public or private sector employees. Some cities will not allow superior officers to join unions representing rank-and-file workers. These restrictions are usually based on the fear that officers would face a conflict of interest if they joined such organizations.

For most chief officers, unionization is a fact of life that must be dealt with. In other cases, the IAFF suggests that management not attempt to oppose unionization. The labor union notes that the law apparently supports public employee organization, which now seems to be a social tide that no one can hold back.

Reasons for Unionization

There are a number of reasons why public safety employees organize unions and why they join them. Charles R. Greer and D. Scott Sink, in a monograph on public sector bargaining in Oklahoma, say that unionization has grown among public safety employees in part as public safety departments have grown. As the departments got larger, employees began to feel more isolated from management and felt the need for more formalized employee relations. There has also been a tendency to equalize the pay of public employees, ignoring the differences in difficulty and danger among different job classifications. The problems caused by inflation and the knowledge that groups in the 1960s and 1970s used civil disobedience to achieve some of their goals have added strong incentive for public safety employees to organize.

The NFPA in its book, *Management in the Fire Service*, notes that there are a number of reasons why employees will seek representation (Figure 4.2). Often they are concerned about their

Figure 4.2 Employees join unions for a number of reasons: to air grievances, to bargain for higher wages, and to safeguard jobs.

jobs. They might worry that there are not adequate safeguards to keep them from getting fired or laid off. They might have complaints that management refuses to listen to or resolve. They might think they have been mistreated. Employees might also be dissatisfied with their wages, salaries, or fringe benefits. There might not be enough communication between employees and their supervisors, which can have at least two effects. In the first place, employees will have little information about their performance. They might know they are performing adequately, because they have received no complaints about their job. They also need to know when they are doing well, and that information never comes down from management if there is a lack of communication. In the second place, employees might feel that their opinions and comments never reach management and that they can have no impact on management decisions and department policy. Employees might also be faced with poor working conditions and discrimination on the job. All of these conditions can encourage employees to join unions.

The IAFF in a course on personnel management for the fire service points out that employees form unions once they recognize they have a "common occupational identification" and come to believe that the company is not doing as well by them as it should. They then use the union to get better wages, salaries, and benefits and to improve their status on the job. They realize, the IAFF contends, that an employee has more clout when organized than standing alone. With a union, they feel they have a better chance to negotiate a good contract and work out employee grievances.

Union Organization

Most full-time career firefighters belong to locals affiliated with the IAFF, which in turn is a member of the AFL-CIO. The IAFF says it differs from most of the industrial and trade unions that belong to the umbrella labor organization. Most unions seem to have quite a contrast between their locals and their national or international structure. The locals are usually organized democratically, with members voting to select their officers, decide their initial positions in collective bargaining negotiations, and ratify contracts. But the national and international organizations, at least in unions representing private sector industry, are likely to have strong control over the locals. The IAFF says it differs from the industrial unions in this respect. It says its international officers help organize locals, resolve local labor disputes and work to improve local working conditions, fire fighting equipment, and department management practices while letting the locals operate independently without strong control from the top. The IAFF also lets fire department supervisors join the union and includes them in the bargaining unit.

CONTRACT NEGOTIATIONS

One of the ways fire chiefs get involved with firefighters' unions is through contract negotiations. These proceed in the public sector much as they do in the private, but there are a number of factors that make the situation in the public sector considerably different.

Joseph E. Benson, writing in the monograph on public sector bargaining in Oklahoma, noted that public sector departments are often the only source of the service they offer in a community. In other words, public sector departments often operate a monopoly. Citizens would be hard pressed to find another way to fight fires if the fire department were on strike. Benson added that public sector services are often vital to the community, which is one of the reasons government initially hesitated to give public sector employees the right to bargain collectively and one reason why public employees have seldom been given the right to strike. Government theoretically represents the rights of all, Benson said, and collective bargaining sometimes forces the government to concede to the demands of a small group, to the detriment of community interests. This usurpation of power could destroy government.

Another difference is that public sector collective bargaining is more likely to involve the courts. In fact, Benson said, increasing court involvement is one of the trends in public sector collective bargaining, and more and more the courts seem to be siding with management in labor disputes. Finally, the law governing collective bargaining in the public sector is not as uniform as law governing the private sector. Some have suggested federal regulation, but most collective bargaining in the public sector is now covered by state law.

In the collective bargaining process, management and union teams have to negotiate an agreement that union members then have to approve. In the public sector the process does not end there. The agreement also has to be ratified by the city council, the state legislature, or whatever body holds ultimate responsibility to the public (Figure 4.3 on next page).

The IAFF also points out that there is a difference in union strategy. In the private sector, union representatives often argue that there is too much of a gap between the company's profits and the workers' wages, salaries, and benefits. Since there is no profit motive in the public sector, that argument will not work for union negotiators. Instead, they often argue that the public sector employees are worth more than they are paid. Nevertheless, the goal is the same: improvement in the employees' situation.

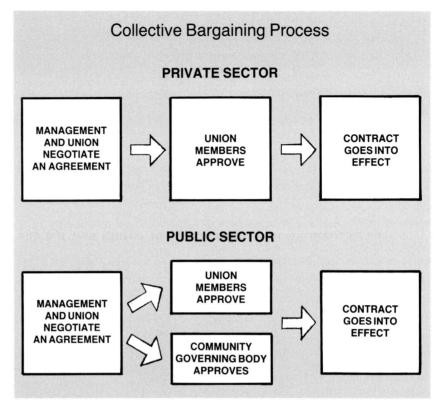

Figure 4.3 For public sector jobs, the agreements reached by management and unions also must be approved by the local or state body which is responsible to the public.

Maintaining Open Communications

The collective bargaining process is basically one of communication. Both sides have to make sure the messages they transmit are the ones received. If not, misunderstanding occurs and the process can break down. Sometimes one side knows what it wants to say, but the message does not come out that way. That is because people are individuals with different attitudes, values, beliefs, biases, and assumptions. Their experiences are different. The same uniqueness sometimes prevents the other side from receiving the message as intended. There can be additional problems on the receiving end. The receivers may be defensive or preoccupied. They may have emotional blocks or hold stereotyped views that get in the way of understanding. Their expectations could prevent them from receiving the message accurately.

The IAFF suggests three ways to get around communications problems. One involves the concept of "quality communications." Negotiators have to think before speaking, finding just the right words and phrases, just the right voice inflections and facial expressions, to make sure the message comes out as intended. The second way is tied to the first. Communicators have to understand their audience. If members of one side know who the message is intended for, they can tailor it to make sure it is received and understood correctly. Finally, the IAFF suggests using a "two-way dialogue." Pass on the message and then use questions to make sure the other side understands.

THE NEGOTIATION PROCESS

The NFPA says that the agreement negotiators work on usually covers five areas:

- Routine clauses, which include constitutional items such as the preamble and purpose of the agreement; terms covered by the agreement; reopening conditions; and amendments.

- Clauses affecting union security. These usually describe the bargaining unit the agreement covers and list the steps by which a union is recognized as an employee representative.

- Clauses describing the rights and prerogatives of management, which recognize that management has the right to decide matters the agreement does not cover.

- Sections that describe how the department will handle employee grievances.

- Sections that list the conditions of employment, describing such areas as wages, salaries and fringe benefits; hours, holidays, vacations and leaves; apprenticeships and training; hiring and firing; safety; and strikes and lockouts.

In negotiations, management and labor are usually represented by teams (Figure 4.4). The IAFF suggests that the management team have a high-ranking personnel officer, a city financial expert or budget analyst, and a lawyer. The fire chief or another upper-level fire department official should be on the city

Figure 4.4 Teams representing management and labor meet to begin the bargaining process.

team, available for consultation and advice. All the other top-level city officials, such as the mayor and the city manager, should also be available.

The bargaining process is similar to buying a used car. Sellers ask for more than they hope to get and buyers offer less than they are prepared to pay. The same thing happens in contract negotiations, and both sides know it. Union representatives ask for more than they are willing to settle for, and management representatives offer less than they are willing to agree to eventually.

Bargaining Preparation

Neither side goes into negotiations without preparation. Gathering information is an ongoing process. Collect data throughout the year. At the same time, keep an eye on how well the current contract is working. Provisions that are not working can become items for negotiations during the next bargaining session. Also, keep on top of any new employee grievances that have cropped up during the contract period. The IAFF also suggests that management work on its attitude toward labor throughout the year. This can affect the tone of negotiations once bargaining sessions begin.

Before entering into contract negotiations, the management team should be aware of a number of things. Members should know the history of previous bargaining sessions with the union. They also should have analyzed the provisions of the current contract and identified any problems involved in its operation. Finally, team members should have learned as much as possible about the members of the union team — their backgrounds, personalities, and negotiating style.

The city team should also have data to support its position. Team members should know what firefighters make in nearby fire departments and what they make in similar departments around the country. The management negotiators should also know how much similar positions pay in private industry. They should know how much current wages and fringe benefits cost the city and how much of an increase the city can afford. The IAFF also suggests that city negotiators be up-to-date on government cost-of-living indexes, but others suggest that some of these indicators, such as the federal Consumer Price Index, might not be appropriate for negotiating new contract provisions. Bureau of Labor statistics studies indicate that the CPI does not do a good enough job of measuring cost-of-living increases to be used alone during contract negotiations. According to The Management Report of the International Association of Fire Chiefs, the CPI puts too much emphasis on housing prices and mortgage rates, and fails to indicate changes in consumer behavior or report levels of consumer satisfaction. The CPI also reflects changes in health care costs, which most fire department contracts cover separately. Thus, the IAFC says, the CPI tends to overestimate increases in consumer costs.

Presenting Proposals

Some suggest that management enter the bargaining session armed with proposals to make to union negotiators, as well as a list of new and continuing demands that management wants to discuss (Figure 4.5). Presenting such a list lets management take the initiative rather than forcing it to react to union proposals. The goal is twofold: enabling management to start negotiations from the same position of strength as the union, and providing a number of items to trade off for union concessions if necessary.

Figure 4.5 Management does not just react to labor demands; the management team should also be prepared to discuss its own proposals.

Scheduling Bargaining Sessions

Bargaining sessions are usually scheduled after the union issues a formal call, typically about 60 days before sessions are expected to begin. Contracts are normally thought to be extended without a formal call for negotiations. Management should be sure to schedule sessions when top negotiators can attend. Once both sides agree on a schedule, management should be sure to follow it rigorously. Otherwise, the union could accuse management of stalling. Be prepared for the negotiations to take time. Each side will want to offer proposals, and each set will have to be studied by the other side. Each side will also need time to offer counterproposals.

Negotiating Contract Issues

The NFPA says there are three categories of real issues involved in contract negotiations. One category involves wages and fringe benefits. Of course union representatives try to get the best possible deal for their members. Management, on the other hand, wants to get by with limiting an increase as much as possible. Both sides recognize that their demands cannot be unrealistic or their positions unyielding. Management knows that a fair package of wages and fringe benefits helps the department attract and retain competent employees. Therefore, good managers know they should not oppose reasonable, legitimate union proposals. Normally, management on this issue tries to start low and bargain up.

Another important issue involves working conditions. Competent managers know that good working conditions also help attract and retain capable people and increase productivity. Good working conditions also help union leaders, who prefer representing happy, satisfied employees who do not take up time with complaints and grievances about working conditions.

A final issue concerns the job security and career advancement of management and union leaders. A lot more than wages and working conditions ride on contract negotiations. Members of the bargaining teams have a lot of personal incentives at stake. Management and union leaders who seem to lose too much during collective bargaining might also lose their jobs. Administrators can be replaced and new union leaders can be elected by the rank and file. If both sides are competent, neither wants to hurt the other so much that one will be replaced. Good managers know that competent union leaders can keep some of the more radical union members in line and head off obviously excessive contract demands. Good union leaders also know that the concessions they can get from incompetent administrators will look attractive to their members over the short term, but big settlements can eventually come back to haunt the union in the form of poor wage and benefit increases later.

WAYS TO HANDLE AN IMPASSE

Despite good-faith negotiations by management and labor, bargaining can hit an impasse, a sticking point over which neither side is willing to compromise (Figure 4.6). There are three ways of getting around an impasse without resorting to a strike: mediation, arbitration, and fact-finding.

Mediation involves bringing in a third, theoretically neutral party to talk with each side and bring out the real issues and concerns that are holding up negotiations. The mediator can clear up misconceptions one side holds about the positions of the other and get both sides talking again in hopes of leading them to a recon-

Figure 4.6 When neither side is willing to compromise on a particular issue, a strike is not the only solution.

ciliation and a contract. The mediator often uses information from other labor disputes around the country to move one or both sides away from unrealistic and untenable positions. The mediator can also get both sides access to high-ranking officials and smooth the way to a settlement. Considering how costly strikes can be, some say that hiring a mediator to resolve an impasse is often cost-effective.

Arbitration, sometimes required by state law or city charter or ordinance, is usually binding on both sides. The process can use a single arbitrator or a panel of three, usually chosen by both sides from a list supplied by a professional organization, such as the American Arbitration Association. Management and labor usually use the so-called "strike-off" procedure to select an arbitrator or an arbitration panel. Each side will alternate to strike a name from a list of professional arbitrators until the required number is left. The arbitrators then hear evidence from both sides in the dispute and come up with a binding solution. Neither management nor labor particularly likes the procedure because it takes the final decision out of their control. Many fire departments are beginning to favor binding arbitration over the possibility of facing a strike and the damage it can do to public confidence.

One form of arbitration some say is gaining popularity is called "final offer" arbitration. In this variation, each side comes up with what are supposed to be its most generous offers on each issue to be resolved. The arbitrator must choose one of the offers on each issue without compromise. The procedure theoretically forces each side to make realistic proposals while coming close to what its final offer on each issue would be.

Fact-finding is similar to arbitration. Arbitrators look at the facts and come up with suggested solutions, but their suggestions are not binding. Some say the procedure identifies facts that can convince city councils and other policy-making bodies to make concessions in return for a settlement. Neither management nor labor is forced to make a serious effort to come up with its strongest offers. The procedure never actually resolves a dispute, and the suggestions might not satisfy either party. Nevertheless, fact-finding and mediation are the most common ways to resolve impasses in the public safety sector.

STRIKE MANAGEMENT

The strike is a last resort the union uses when it sees no other way around an impasse or wants to pressure management to grant concessions (Figure 4.7). Strikes by public employees are

Figure 4.7 Although illegal, a strike may be used as the final tool to force management to grant concessions.

usually against the law. Rhyne notes that even without the statutory ban, strikes are still prohibited by common law, which looks upon them as examples of anarchy and conspiracy against the government. Some states give public employees a limited right to strike, but public safety officers are usually excluded. Some contracts have no-strike clauses. The IAFF used one for 50 years until 1968, when the union dropped it as evidence of its increasing militancy. Bans against strikes have had little ability to prevent them, although their numbers have decreased in recent years (Table 4.1).

TABLE 4.1
FIREFIGHTER STRIKES 1976-1981

Year	Firefighters Only		Combined Police/Firefighter	
	Number of Strikes	Personnel Involved	Number of Strikes	Personnel Involved
76	12	1600	3	1200
77	15	2600	5	500
78	15	4800	4	500
79	16	1100	4	3500
80	9	5800	6	2300
81	3	1200	1	136

Source: U.S. Bureau of Labor Statistics

It is estimated that strikes by public safety officers last an average of three to seven days, and they can be brought on by a number of issues: problems with wages and hours, disputes over job benefits, unhappiness about pay parity, and unhappiness about the way management handles such union issues as recognizing a union or defining the bargaining unit. Officers will sometimes go on strike to protest demotions and layoffs. But one central issue will usually be the primary cause of the strike, and management leaders can sometimes head off strikes if they react to the first indications of employee discontent by trying to identify the primary issue involved. Then they should move quickly to handle it. Later they can look for other causes of dissatisfaction and resolve them before they lead to further unrest.

Contingency Plans

If the department cannot avoid a strike, management should have a strike contingency plan ready to put into operation (Figure 4.8 on next page). Some say the plan should have two basic functions: to let higher administrators and supervisory personnel know what their strike-related activities will be and to make sure the department can continue to provide emergency service.

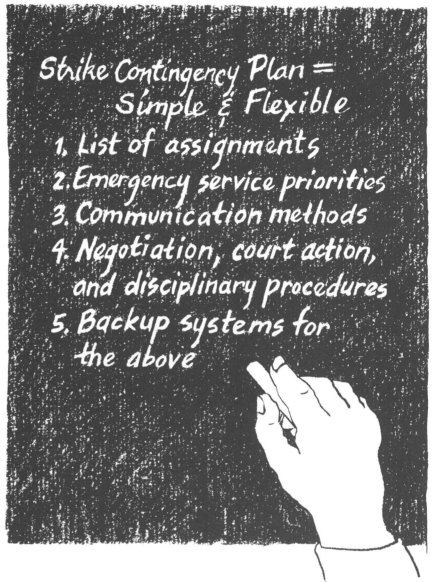

Figure 4.8 A department needs a strike contingency plan in order to continue to provide emergency services.

The strike contingency plan should be drawn up in advance and updated annually. This lets the department stake out its position and tells the public and department employees that the department is ready to operate as normally as possible should firefighters go on strike. The plan should be simple and flexible, and should include the following items:

- A list of assignments so supervisory personnel will know who does what

- A list in priority order of what emergency services the department intends to provide

- A method of communicating with department personnel — both off and on strike — and the public

- A description of the administrative machinery that man-

agement can use for negotiation, court action, and discipline
- A description of back-up systems for each of the above

Only a small group of officers should draw up the plan in order to keep it flexible and secure. One of the planners should be the fire chief, and all members of the planning group should know fire science and fire administration.

The department can also set up a strike coordinating committee to handle a number of tasks like those listed below:

- Reviewing the strike plan and keeping it up-to-date
- Naming a strike coordinator to deal with emergency calls
- Pulling together information about the strike for distribution and watching how the strike plan operates
- Handling discipline and analyzing the legal ramifications of any action the department has to take
- Looking at the departmental communications system and organizing a communications center
- Reviewing the department's personnel policies and regulations to see how firefighters might respond to departmental action and discipline
- Deciding what priority the department should put on different emergency services and planning how to get personnel to substitute for strikers
- Setting up security to protect department equipment, facilities, and nonstriking personnel

Anticipating Employee Actions

Before the department puts the plan into effect, administrators should see what kind of job action is under way to determine the department's response. There are a number of tactics unhappy workers can employ without going out on a full-fledged strike. There are walkouts and "sickouts." Employees can report to work and then try to disrupt departmental activities. They can try a form of sitdown, where they show up for roll call and then refuse to move out to their duty positions or assignments. Other groups have engaged in "rule-book" slowdowns, when employees follow all the rules with agonizing slowness. They conform to the letter of the law and refuse to take some of the legal shortcuts that contribute to speed and efficiency.

Administrators usually get hints about when a job action is about to come off, especially if they are in negotiations with the union. Watch the tone of union representatives during bargaining sessions and monitor their comments to the news media for clues. Pay attention to visits from national union officials. If the department does not have a strike contingency plan, administrators should start checking out the legal implications, getting a plan together and asking other cities for advice as soon as conditions seem to favor a strike. Regardless, put the plan into operation as soon as employees give formal strike notice.

The following are some questions that management should consider in the early stages of a job action:

- When should administrators implement the contract's no-strike clause, if any, or call in statutory provisions that ban and punish strikes?
- Does the department have the right to impose work orders?
- Can the department ask the courts for restraining orders and injunctions?
- How should the department handle the strike? Should it work internally? Go through the city council or the mayor's office? Use the courts?

No matter how the department answers these questions, its main goals should be to continue emergency service during the strike and to end the strike as quickly as possible.

Public Relations and Communications

During the first hours and days of the strike, departmental concerns will fall into three primary categories: public relations and communications, internal procedures, and a variety of management tasks and responsibilities.

The aim of the department's public relations during the strike is to get the public on the department's side. Public support for the department tends to result in short strikes resolved in the department's favor. The department also needs a way to communicate with top-level city and department officials and with rank-and-file employees, whether on the job or on the picket lines.

A well-organized communications center can handle both tasks. Rhyne suggests that the department assign a special staff composed of a communications director, a public relations liaison, and at least one full-time secretary to operate the center. The director is responsible for clearing all publicity, and the liaison makes all statements for the department.

All communications with the public should try to convince citizens that the department can still protect the community even while firefighters are on strike, and then get public support behind the department. Management could start with a press conference to assure the public that the city can still provide fire protection and to explain the department's position on the strike. Then the department can issue position papers and press releases. Administrators can speak to citizens' groups and grant newspaper and broadcast interviews. One administrator even debated a union leader on television during a strike.

The department also has to be ready to counter union publicity moves. Some strikers have tried to capture public support through sympathy strikes. The department should move quickly to seek pledges from other groups of city employees to stay on the job. Act first so that the union does not put the department on the defensive.

Recognize that the news media is a powerful tool for getting the public on the department's side (Figure 4.9). Although a lot of people complain about the quality of their newspapers and the shallowness of broadcast news, they often believe what they read, see, and hear — and they act on it. The department's task is to get accurate, believable information to reporters. Be sure the communications center is equipped with enough typewriters and telephones to accommodate reporters. Make sure departmental leaders or the public relations liaison are available for interviews. Reporters are under pressure from their editors and news directors to get a story by deadline, and if department representatives are not available, reporters will go with the best information they can get, even if that makes the story one-sided. Reporters believe they have done their duty if they have tried to contact all sides on an

Figure 4.9 Utilizing the news media can be a powerful tool in dealing with a strike situation.

issue. If they are unsuccessful, they will report that one group could not be reached for comment and leave it at that. It also helps to know when newspapers and broadcast stations have set their deadlines so the department can issue information for maximum impact.

Sometimes a strike will be so important that the national news media will send reporters to cover it. Department leaders should realize that while national media attention is flattering, it has little effect on local public opinion. Therefore, if the department caters to any group of reporters, it should cater to local ones. They will do a better job of marshaling public support behind the department.

The following information is typical of what the department might want to put out:

- Descriptions of how substitute workers are handling the job
- Pledges of support from other segments of the community
- The progress of negotiations to settle the strike and produce a contract

The department also has to handle internal communications. Be sure that each employee, striker and nonstriker alike, has a detailed statement of the department's position on the strike. It is better for them to get it from the department than to rely on the abbreviated version in the news media. While detailing the department's position, avoid criticizing union leaders and other individuals. The courts could consider this an unfair labor practice, and it only serves to make the situation more emotional, volatile, and difficult to resolve.

INTERNAL PROCEDURES

The department's second major concern during the early stages of the strike is with internal procedures (Figure 4.10). How does the department react toward its employees when they turn in a strike notice? Rhyne gives several suggestions. First, tell union members and their leaders what the department's position will be. Management can ask the union president not to encourage picketing, which can hamstring operations and encourage other employee groups to stay away from facilities surrounded by picket lines. Spell out the department's position on disciplining strikers. Be sure they understand that the job action is illegal, if such is the case, and what can happen to them if they persist. Do not assume that everyone is joining in the job action. Call roll and note the names of those who do not show up or who refuse to take an assignment. Let the chief or higher-level administrators, rather than the normal supervisors, decide what disciplinary action to take. This should ensure evenhanded treatment.

Figure 4.10 Be sure all employees know what actions the department intends to take if there is a strike.

Management Tasks and Responsibilities

Department leaders will have many tasks and responsibilities to handle during the first hours and days of a strike. Meet with other fire department leaders and city officials to decide how to implement the strike contingency plan. The first priority now is to settle the strike. Management negotiators can worry about handling employee grievances later. Make sure union leaders are available for talks around the clock. Management should contact other groups of city employees and encourage them to stay on the job and to pledge not to interfere. Find out what legal action the city can take, and contact insurance representatives for their help in preventing property damage. Department leaders also have to decide what to do with employees who want to continue working. Probably most important, the department should develop its own plans to end the strike rather than waiting to react to union suggestions.

DETERMINING WHAT SERVICES TO PROVIDE

During the strike, the department has three operational questions to answer:

- What priority should the department give to various services? Which ones have to be maintained, and which ones can be suspended?
- How does the department continue to provide emergency services during the strike?
- What security does the department need and how should it provide it?

During the strike, the department will have some hard decisions to make about what services to continue and what to cut during the strike. How well can a smaller and possibly inexperienced work force handle the job? What would be the start-up costs of a suspended service, and what political consequences could the suspension have? Who will the public hold responsible for reduced services, the department or the striking firefighters and their union? Administrators should place the highest priorities on emergency services such as fire suppression and rescue. In some strikes, firefighters have continued to respond to emergency calls while refusing to perform such other duties as fire inspection, training, and station maintenance. If such is the case, the city can let substitutes handle the nonemergency duties.

The department's main goals should be to provide enough service to keep up community support while avoiding a court suit. There have been instances in which citizens have filed civil cases against cities and departments to collect damages for injuries received during a strike. The IAFC in its *Management Report* described two cases arising out of the 1980 Chicago firefighters' strike in which the plaintiffs tried to hold striking firefighters and their union responsible for fire deaths and property damage. The IAFC said that many lawyers specializing in labor relations believe that striking firefighters can be held liable for deaths that occur during illegal strikes if the plaintiffs can show that the shortage of personnel contributed to the deaths, and if the union could have reasonably foreseen that the strike could have resulted in such incidents. The Chicago cases had not been adjudicated by mid-1983.

CONTINUATION OF EMERGENCY SERVICES

Among its own employees, the department will have two groups who will want to remain on the job. One group will be composed of supervisors, who should participate in an ongoing program to train them how to fill in for rank-and-file firefighters in case of a strike. The other group will be composed of nonstrikers who want to stay on the job. Management has to decide whether the benefits of keeping them on duty will offset the conflict that can develop between them and their striking colleagues. If the department refuses to let them work, the union can accuse it of a lockout. If the department keeps them on the job, it should warn them that the union or its members might resort to reprisals or harrassment. The department should make every effort to protect nonstrikers.

The department can also get help from the National Guard. Departments should know the procedure for calling out the guard and ask the governor to put the guard on alert as soon as a strike seems probable. Be sure the governor mobilizes enough troops to handle the job; if not, be prepared to go elsewhere for help. Recognize that the National Guard is better at law enforcement than fire fighting, which is not usually a part of the troops' training. Instead of the National Guard, the department might prefer asking the state forest service for rangers who at least are trained to handle forest fires. The department could ask the military for fire fighting personnel. Some departments have set up and trained volunteer brigades in case of a strike. Others have tried to subcontract the work with private organizations or have hired temporary or permanent employees to replace strikers. Some departments have signed mutual aid pacts with departments in other cities to supply outside firefighters when local employees go on strike. Such pacts might not be reliable if the out-of-town firefighters are unionized and sympathize with the strikers. However the department handles the personnel problem, officials should make sure that equipment and supplies are stockpiled and that room and board is provided for the new workers. Management should also schedule regular rest periods to keep personnel working overtime fit and alert.

PROVISION OF SECURITY

Finally, the department must provide security for equipment, facilities, and nonstriking personnel during the strike (Figure 4.11 on next page). Most striking public safety officers abide by the law, but there have been instances of misconduct. Cities should be prepared to head them off or handle them when they occur. This issue is one of the toughest the fire department will have to face. Most departments in the public sector can rely on the police and fire departments for security when employees go on strike, but fire departments do not have this option. When firefighters are on strike, sometimes police officers will refuse to help out of sympathy. Here are some of the actions department leaders should take to secure their employees, equipment, facilities, and the public from harm:

- Check security arrangements for public buildings and identify the ones susceptible to sabotage.
- Determine which buildings should stay open and which should be closed.
- Set up any electronic surveillance that has to be used.

Figure 4.11 Plans must be made to provide as secure a working environment as possible for management and nonstriking employees.

- Keep an eye on picket lines for potential misconduct or violence.
- Collect keys and vehicles in a central location.
- Consider limiting access to department facilities.
- Decide whether to provide transportation and guards for substitute workers.

Throughout the strike, management should work to end it and get the strikers back on the job.

Settling the Strike

One way to resolve the strike is through negotiation and settlement. The tactic here is to concede just enough to get the strikers back to work. Conciliation and compromise will often work better than hard-nosed resistance. The fire chief's role here is that of advisor, and the chief should be available when needed for consultation. Whoever conducts the negotiations should be prompt and businesslike. The union representative can easily characterize any delays as bad-faith negotiating on management's part. The city will probably have to grant some kind of pay increase to get the firefighters back to work. It will also have to make other decisions leading into negotiations to end the strike. Should the city even negotiate with the union or should it withdraw its recognition of the organization as a bargaining agent for its employees? Should the city grant amnesty to strikers to get them back to work? If so, should it be to all or should it be just to those who have acted legally and who return to work by whatever deadline the city sets? If a city grants amnesty to those who have participated in an illegal job action, it could encourage other municipal employees to strike later.

Cities can also turn to the courts for help in resolving a strike. Most such legal action is based on the claim that the strike is illegal and that the court should restrain or enjoin strikers from breaking the law. In some cases, courts have enjoined strikers even when the city has not made such a claim. Cities also typically contend that the strike damages the city's ability to deliver essential services to its citizens. Cities do not have to rely on the typical irreparable harm standard necessary in most successful equity cases. Often plaintiffs can expect to get restraining orders or injunctions only if they can prove they will suffer irreparable harm if the defendants are not restrained or enjoined. In cases involving strikes by public safety officers, the courts assume that irreparable harm has in fact occurred. Neither do cities have to claim a breach of the peace.

Management will usually want to file the case in state court, and the city should be ready for any delaying tactics the union might employ. Sometimes the union will challenge the jurisdiction of the state court and try to get the case into the federal system. Union leaders can also ask for continuances (postponements) to prepare paperwork. Management should realize that any delay only works in favor of the union, giving its leaders more time to get the strike under way. Cities have fought such dilatory tactics by claiming that the safety of the public demands a speedy hearing of the case.

The courts will not always provide a successful solution, even if the city wins its case and the judge orders strikers back to work. Some union leaders and strikers, their attitudes hardened by a restraining order or an injunction, have defied back-to-work orders. Such court orders have sometimes made strikers even more attached to their unions. The courts can punish defiant workers, their union leaders, and their union with jail sentences and fines after finding them in contempt. This can create public sympathy for the strikers and makes the job of conciliation after the strike much harder, if not impossible. Before resorting to the courts, management should think about the following questions:

- What influence does the union hold over strikers?
- How does the size of the work force compare to the number of strikers?
- How effectively are substitute personnel doing their jobs?
- How are the negotiations going?
- What would be the political ramifications and consequences of going to court?
- Would the judge want to get heavily involved in department administration?
- How willing might the strikers be to return to work after missing a paycheck?

Final Considerations

Once the union agrees to end the strike and the strikers are ready to return to work, the department has to face the tasks of reorganization and reconciliation. The department should formally notify all employees that the strike is over. Administrators should check the records they kept during the strike to see if anyone should be punished for misconduct, assuming the agreement does not include an amnesty clause. Department officials should also go over the strike contingency plan to see how well it worked and whether it should be modified.

The question of discipline is a difficult one. Some states require it by law. Discipline can cause more problems than it solves, especially if complaints about discipline led to the strike in the first place. Any discipline the department undertakes has to be fair and evenhanded to avoid future complaints. Be sure to satisfy all the requirements of due process and follow all the legal guidelines. Department officials need to have proof of misconduct before taking any action. For example, firefighters claiming sick leave may be required to provide a doctor's excuse.

Discipline can range all the way from verbal reprimands to dismissal, although departments seldom resort to the latter. Dismissal is hard to handle because of the legal requirements of the process and the problems of hiring qualified replacements quickly. Discipline often is a better solution than contempt of court to punish strikers. It lets the department keep control of the situation and does not have to have the threatening overtones of a judicial proceeding. The department has the flexibility to match the punishment with the offense, and no one is in danger of going to jail. Some believe that handling discipline internally avoids publicity, but leaks from a variety of sources can quickly get out the word of internal discipline to the news media.

Some citizens have amended their city charters to specify the types of punishment strikers should receive. Realize that any sanction is detrimental to employees, because notice of the sanction almost always winds up in their personnel files. And a sanction could force the dismissal of probationary employees. If that is the case, the city will have problems finding qualified people and finding the money to hire them and train them. The city can rehire those it fired without having to extend the same terms and conditions of employment. Regardless, having to fill vacancies will only delay getting the situation back to normal.

Fines and sanctions against the union might be more effective in heading off future problems than any action the department might take against an individual. The department could withdraw the union's recognition as a collective bargaining agent and its dues checkoff privilege. The department could also sue the union for damages incurred during the strike and seek fines against the organization and its leaders. The union might also be forced to pay court costs for members sued in connection with the strike.

Information Management

5

NFPA STANDARD 1021
STANDARD FOR FIRE OFFICER
PROFESSIONAL QUALIFICATIONS

Fire Officer VI

7-2 Management Information Systems.

7-2.1 The Fire Officer VI, given a simulated or actual record system shall:
- (a) demonstrate knowledge of how to analyze the records and data
- (b) demonstrate knowledge of how to interpret the records and data and determine validity
- (c) demonstrate knowledge of how to evaluate the data for the purpose of recommending improvements.

7-2.2 The Fire Officer VI, given a summary of the goals and objectives of data processing services and systems within the jurisdiction and a summary of the components and operational principles of various types of data processing equipment shall:
- (a) identify the capabilities and limitations of electronic data processing equipment
- (b) identify the components and functions of electronic data processing hardware
- (c) identify and define the functions of data input, sorting, processing, fielding, retrieval and control.

7-2.3 The Fire Officer VI shall demonstrate knowledge of how to direct the development, maintenance, and evaluation of the department record system to attain completeness and accuracy.

7-2.4 The Fire Officer VI shall demonstrate knowledge of the principles involved in the acquisition, implementation and retrieval of information by data processing as it applies to the record and budgetary processes and response patterns in the fire service.*

The above NFPA standards are addressed from a general management perspective.

*Reprinted with permission from NFPA No. 1021, *Standard for Fire Officer Professional Qualifications*. Copyright 1983, National Fire Protection Association, Boston, MA.

Chapter 5
Information Management

Information management is a skill. It consists of planning, controlling, generating, processing, preserving, evaluating, and disposing of both operational and management information.

Generally speaking, firefighters and officers regard recordkeeping and reports as "busywork" and look at it as a necessary evil. This is unfortunate because lack of information is one of the main reasons that the fire service is having difficulties maintaining adequate staffing and budgets.

Two of the most obvious reasons why we lack a good information network in the fire service are time and money. Less obvious is the fact that we have not developed enough expertise in managing information systems. True, we have forms — many of them — but using information to aid in decision making has had a lower priority because emphasis has traditionally been placed on emergency operations.

This attitude is changing, however. Master planning, tax revolt, and the new breed of firefighters are pushing toward the recognition of information as a valuable tool. Good information that allows for internal control of the environment is a source of power. Those individuals within local government who control information systems, especially automated ones, will gain power at the expense of those who do not possess the information.

Situation: The fire administrator stands at the end of a long table, defending the department's budget request, as city council members rock back and forth in their swivel chairs and make notes. Two of the five council members, elected last April on pledges to overburdened taxpayers to cut city expenses, ask increasingly harder questions about some of the administrator's proposed expenditures for the upcoming fiscal year. The city manager leans forward. The newspaper reporter stops doodling and starts writing down quotes.

Situation: A motorist driving through a recently-opened industrial park sees smoke coming from a warehouse and dials the 911 emergency number at 9:14 p.m. to alert the fire department (Figure 5.1). The dispatcher, following departmental procedure, sends three companies to the scene, and the first to arrive reports that firefighters are laying a line.

Figure 5.1 Firefighters who already know the safest and most efficient way to respond to an area will do a better job protecting property and lives.

While the situations are fiction, the questions they raise are quite real. Does the administrator have the facts and figures to parry the questions of council members about the proposed budget? Do the firefighters responding to the industrial park alarm have up-to-date information about the best way to reach the warehouse, the location of hydrants, the construction of the structure, and hazardous materials that might be stored in the building? If the answers are yes, the department stands a good chance of getting what it needs to provide adequate fire service to the community, and the firefighters can hold property loss to a minimum with no more than the expected risks. Favorable outcomes in both situations depend on having the right information in a good record-keeping system, and that depends on the department's ability to gather information, store it, analyze it, and retrieve it when needed.

The fire department, of course, needs good records for other reasons besides justifying budget requests and making decisions on the fireground. Good records are necessary to run the department on a day-to-day basis, and to keep up with equipment maintenance and personnel matters. They are a must for planning for the long run, too, to establish goals for carrying out the department's mandate from the community. Once new programs have been in operation awhile, good records are required to determine if the programs are doing what they were set up to accomplish. Departments can use a good record-keeping system to devise or revise fire codes, establish priorities for research, regulate new products that seem to have been involved in an unusual number of fires, and back up management positions in labor negotiations. Finally, good records are useful in telling the department's story to the community, reporting accomplishments, and warning citizens about dangers to their lives and property.

RECORD SYSTEMS

Beginning in the late 1880s, and especially during the first two decades of the 20th century, the insurance industry prompted the development of many of the basic record-keeping procedures we use today. Such professional associations as the National Fire Protection Association, International Association of Fire Chiefs, International Association of Fire Fighters, and the International Fire Service Training Association have had a large impact on the growth and development of recommended documentation forms.

Not surprisingly, a system that has grown in such a piecemeal fashion over the last eighty or ninety years has developed some problems. For the most part, fire department records, reports, and other documents are not providing relevant information; neither are they being used to support the development of a total fire protection system. In many cases, the paperwork that is generated is more of a defense mechanism than a management tool.

What is needed is expertise in *managing* the informational needs of the fire service, and this means managing the record-keeping system. Such a system should include the following:

- A comprehensive plan for the entire document system, thereby eliminating gaps in the department's knowledge base.

- A conscious effort to use the system of records and reports for making decisions or changing programs.

- Data collection methods that enable the department to engage in straightforward and original thinking regarding its own needs.

- Cooperation and concern for the development of broad-based systems that encompass regional, state, and federal levels. This is needed so that a more scientific analysis can be performed on the true nature of our nation's fire problem.

The simpler a system, the more desirable it is. While this may seem an unusual approach in view of the importance of information systems, it is true. In some respects, the larger an agency is, the simpler its system should be because of the volume of data. However, the opposite often occurs. Many small agencies have only the sketchiest of systems, while the larger agencies have more forms than people to analyze them.

An efficient reporting system will generate mounds of data that can easily overwhelm fire department personnel if it is not well organized (Figure 5.2). All members must understand the system's components, why certain documentation is needed, the proper technique of utilizing the data to make decisions, and the methods used to implement changes in the system to keep it current.

Figure 5.2 Information which cannot be managed soon becomes useless.

Record-keeping systems will vary from department to department, but each department will have the same goal in mind: defining the scope of the community's fire problem. Some departments keep their information on paper in a variety of reports. Others have computerized their files and microfilmed the originals for back-up, but most record the same basic information in the same basic format. Among the most common are records and reports on incidents, fire investigations, building inspections, pre-fire planning hazards, dispatch, equipment, personnel, and budgeting and finances. In addition, some departments may use the records of other local agencies for information on building construction and occupancy (from the city planning department and tax assessor-collector), and socioeconomic data from planning, social service agencies, and the U.S. Census Bureau.

REPORT FILING SYSTEMS

Information in a record-keeping system will have little value if personnel cannot find it quickly. Most departments try to institute a logical, orderly filing system, keeping records in alphabetical order by subject category or in numerical order according to a classification system. For example, departments could use the coding numbers found in NFPA 901, *Fire Reporting System,* to keep their files in order. Some departments, such as the Orlando, Fla., Fire Department, have devised their own systems. The training divison in Orlando has used a six-digit numbering system to provide ready access to its records. The first three digits represent the office to which the file belongs, such as the in-service training section or the special and recruit training section. The fourth digit, which follows a hyphen, identifies a major functional area, such as research and development or training records. The last two digits are subfunctional category numbers assigned by the office supervisor. The division has arranged records with the same file number alphabetically.

The filing system would not be complete without cross-referencing and tenure systems. Cross-referencing by numerical code or subject matter will help guide personnel to related information in the system that could be useful. A tenure system will help personnel update records and weed out those the department no longer uses. In most systems, incident reports are permanent records. Legal counsel can give advice about how long to keep other kinds of records. The Orlando training division system referred to above indicates record tenure by the way labels are placed on the file folder: a label on the left identifies a document that needs continual updating, that may be replaced by more recent information, or that will become obsolete; a label on the right identifies a permanent record. Departments should review their files at least once a year to remove or update files and to determine if the system should be revised.

Computer Storage

Many of the nation's fire departments are switching to the computer to keep track of the data they need and to organize it in meaningful ways. Often, the computer hardware — the data storage and processing units — are owned or leased by the city and used by all city departments. Departments can purchase the software — the computer programs that manipulate data — or have city computer experts write the department's own programs. Fire Protection Publications now has software available. Contact Customer Services for complete information.

Before purchasing or designing their own programs, departments should give some collective thought to answering the following questions:

- Who will be using the information? Some people are not as quick at interpreting the standard computer printout as those with more training, and those with less experience will need the information in a format that is easier to understand.

- How will the information be obtained from the computer? Do fire department personnel need direct access to the computer on a video display terminal, or will they need the information only after it has been printed and analyzed?

- How often will the information be needed, and how quickly? Some information, such as the disposition of firefighters and equipment, should be available immediately, but other information might not be needed until the administrator starts work on a preliminary budget.

- How accurate should the information be? Information can be correct within a few percentage points to get a good idea of a trend, but budget figures have to be accurate to the penny. Of course, to be accurate coming out of the computer, the information has to be accurate going in, and this is a function of training and the emphasis administrators place on accuracy among their subordinates. Fire administrators should avoid the attitude that "this is close enough for government work."

Statistical Analysis

Once the system is in place, the chief officer's next concern should be thorough, unbiased analysis. The computer can spill out thousands of words and numbers in less than a minute, but how should they be put together?

One of the most useful techniques is to organize the data in summary form. Computer programmers are experts in devising formats to present data in ways that can be understood easily. For example, they can come up with tables that put data in different categories. Then it is a matter of seeing where most of the cases seem to fall to determine where the problems are. If chief officers can tell from a table that more alarms have come in on the first shift than on any other, then they can decide to increase staffing levels during the busiest period of the day (Figure 5.3). One fire chief in Pittsfield, Mass., learned from computer-produced data that a majority of the structures involved in a fire during a specific period had no smoke detectors. The chief was able to have the fire prevention division contact the owners and/or occupants and suggest that they buy and install the devices.

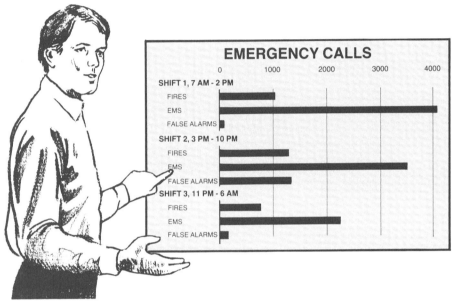

Figure 5.3 Does the department respond to more calls during certain shifts? Grouping numbers by categories enables the chief to see if staffing needs to be increased.

Sometimes the raw figures will not be enough for a valid analysis. Before any decisions are based on the data, the actual numbers should be converted to rates per capita and percentages — statistics that can help fire department analysts identify trends. For example, it is hard to get an idea of the comparative fire problem in two fire districts when the chief officer knows only that last year there were 172 fires in Fire District 1 (population 60,000) compared with 154 fires in Fire District 2 (population 55,000). If those raw figures were converted into rates per 1,000 persons, the chief officer would discover that the fire problem is virtually the same in the two districts, 2.9 fires per 1,000 persons in District 1 and 2.8 fires per 1,000 persons in District 2. The rate, in this case, is a better tool for planning than the raw numbers.

Consider another example. Investigation showed that 70 out of 178 firefighter injuries one year resulted from improper facemask seals. This sounds serious, but is it? Converting the raw numbers into a percentage, 39 percent, confirms the seriousness of the problem and points to a need for change in the training program. Not all figures are so obvious, and using percentages can help identify and clarify a trend.

However, statistics have to be analyzed thoroughly or they can mislead. Mark Twain once wrote that there are three kinds of lies: "lies, damned lies, and statistics." People have used and abused statistics to prove anything in their interest. Here are some things to beware of in statistical analysis:

- Drawing conclusions from a small number of cases. If a city suffered five fire deaths one year and six the next, it would be misleading to call for a budget increase because of a 20 percent hike in the fire fatality rate. The extra death might have been due to chance rather than a deteriorating fire department. Percentages usually have little meaning if they are based on fewer than 100 cases. Analysts can always broaden categories or extend their studies to cover more time to produce figures that will make their percentages meaningful. For example, instead of considering the number of residential fires resulting in losses between $25,000 and $50,000, analysts may have to look at losses between $25,000 and $75,000 to get enough cases for a valid analysis.

- Putting too much faith in statistical "significance." There are several statistical tests, based on the theory of probability, designed to tell researchers whether a numerical difference between two sets of data is due to chance fluctuation or is a "real" difference. It is easy to find statistical "significance" using some of the tests if the number of cases is large enough. Fire administrators should remember that statistical significance says nothing about the difference being important in terms of life, property, or money.

- Introducing bias into the data. This will often come about because of inaccurate data collection, or data collectors not knowing enough about the system of categories they are using to classify data. Report writers have to know the coding system the fire department uses, and those who input data should know how to get data into the system accurately. To catch mistakes, supervisors can check over written reports for completeness, and computers can be programmed to check on the reasonableness of the data (Figure 5.4). For example, a report on a fire at a luxury apartment, where firefighters labored four hours before putting it out, stated the loss totaled $400. Did the report writer mean $400,000? When analysts come up with suspicious results after using data in the system, it is a clue that some of the data might be incorrect. Another bias to

Figure 5.4 Department staffers must be familiar with the data collection system. They should also check reports for completeness and consistency.

watch for: poorly trained investigators have a tendency to find what they expect to find.

- Assuming "unknown" causes are distributed proportionally over the "known" causes. Just because smoking in bed caused 40 percent of the fires for which causes are known does not mean that smoking in bed caused 40 percent of the fires for which investigators could not establish cause.

Information Applications

The uses to which computer-stored data can be put are many. The Pittsfield, Mass., department categorized fires by fire districts to learn where its most serious fire suppression problem was. Coming up with figures showing the number of fires by property use, place of origin, equipment, and forms of heat involved gave the department ideas on where to concentrate its fire protection efforts. Other figures indicated how efficient firefighters had been using different methods to extinguish fires.

The New Haven, Conn., Fire Department has used computer-manipulated data in the fight against arson. Investigators there, for example, could determine if an individual had been linked in some way to more than one arson case or suspicious fire. The department has also worked to come up with arson "indicators," characteristics that seem common in deliberately set fires, and had plans to program the computer to alert the department when a fire had a "trigger" indicator.

The Wichita, Kans., department has programmed its computer to put information on maps that show the relative number of alarms received in each census tract and locations of different types of incidents. The Wichita computer will also print out lists of buildings due for another fire inspection.

Departments have also devised ingenious ways to get the information to firefighters in the field. Fire apparatus in Upper Saddle River, N.J., carried typed cards in a rotary file to have on hand information about how to get to different addresses in the city and where to find water once firefighters arrived. Volunteer firefighters in Little Ferry, N.J., had similar information available on two-minute tapes they could play in the cab of their truck while responding to an alarm. Fire departments in Anaheim, Calif., and Nassau County, New York, have used microfiche to brief firefighters as they drive to answer an alarm. In the New York system, the fire district dispatcher keys in the name of the town and then the first letter in the name of the street on which the incident was reported (Figure 5.5). In five seconds, the computer generates an alphabetical list of all streets in the town beginning with that letter, and the dispatcher can get information from the corresponding microfiche file to radio to responding firefighters. New York City has transmitted information from its microfilmed building hazards file in much the same way. Trucks in a Canadian department carry microfiche files and readers in the cab. The Winslow, Ariz., Fire Department puts pre-fire data on transparencies that fire officers can examine on a viewer attached in the cab of a pumper. The transparencies include building floor plans and typed information about hazards that might be on the premises.

INFORMATION TRANSFERS

Fire administrators must come up with more formal, permanent ways to transmit information to department personnel, city administrators, and elected officials. The department's historical experience as reflected in the data collected in its records can be used to issue statements on standard operating procedure for specific situations. These and other policy pronouncements, along with less important information, can be distributed to department personnel through bulletins and newsletters. The public information and education division of the Phoenix, Ariz., Fire Department periodically issues a newsletter called "Phoenix Fire-

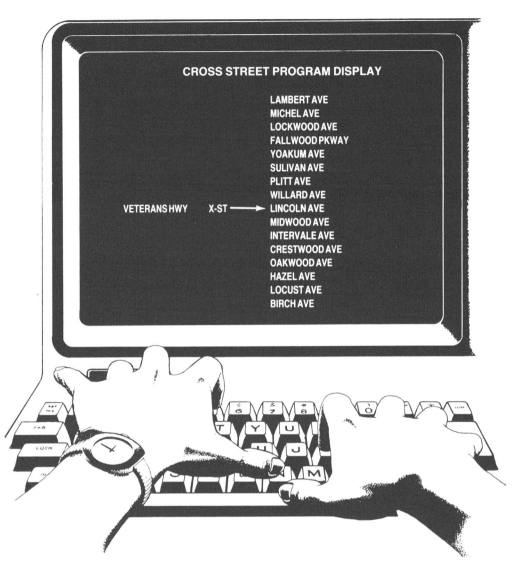

Figure 5.5 Computers can give quick information on street names, cross streets, and routes to the scene of an alarm. *Courtesy of Statustronics Corporation.*

works." The newsletter not only passes along information the chief wants personnel to know, but also tries to answer firefighters' questions and confirm or deny rumors current in the field. Firefighters can dial a telephone number and get a recorded message to pass on their questions, gripes, or rumors to management. Staffers research the questions and investigate the complaints and rumors for publication in the upcoming issue. Personnel and management have found the setup mutually beneficial: firefighters have a chance to blow off steam, and department officials have a chance to learn some of the concerns of their employees and take care of any serious morale problems.

One typical method of formally transmitting information is through the department's annual report. One experienced fire chief listed several suggestions for putting out an effective annual report:

- Determine who will read the report.

- Describe what happened during the reporting period and predict what will happen during the next, along with suggestions for what the department can do about anticipated problems. This section should be strongly backed by hard facts, including statistics comparing what happened during the reporting period with the department's historical experience. This should include information on fire locations, causes, losses, and casualties. The report should also deal with fire prevention, training, equipment maintenance, and finances, although it is seldom necessary to print a detailed budget.
- Make the report readable and attractive.

This last point has been dealt with by communications management consultants. One team suggested that the annual report be as colorful as the business, giving the reader a sense of participation in the community's fire service efforts. Fire administrators can start with an eye-catching cover and continue the effort on the inside with attractive makeup and design and full-color photographs of the department in action. Decrease the cost by writing concisely and printing the report on both sides of the page. Increase the report's usefulness by providing an index. Finally, the consultants suggested, make the report part of the department's public relations effort by sending it to community opinion leaders whose names might not be on the typical municipal distribution list.

PUBLIC RELATIONS AND INFORMATION

A major portion of the fire administrator's job as information manager is getting information to the public. How can citizens protect themselves from fire? How can they react to fire when it occurs to ensure minimum injury to life and property? How can they help their fire department protect them? The department can deal with the public directly through public relations campaigns or indirectly through local newspapers and broadcast stations.

Public Relations

A good public image is obviously helpful to a fire department at budget time, but image-making should not be the primary goal of a departmental public relations campaign. Of more importance is getting information to the public that can save lives and property and help the fire department do its job. The National Fire Protection Association lists several things that should be included in a department's PR strategy:

- Make public relations an ongoing project that provides a continuing supply of information. Getting the public to expect a steady stream of information from its fire department makes citizens more receptive to the messages.

- Time public relations efforts to specific seasons of the year and events. For example, many fire departments prepare promotional and information campaigns to reach their peak just before the Fourth of July or during Fire Prevention Week (Figure 5.6).
- Use news accounts of fire service activity as a peg to provide the public with more fire protection advice.

Figure 5.6 Holidays such as Christmas or the Fourth of July are an opportunity to get safety messages to the public.

Many departments have developed creative ways of getting their messages to the public. More and more departments are beginning to see the potential for public education in public service channels on cable television. Departments in Modesto, Calif.; North Syracuse, N.Y.; Salem, Ore.; and St. Charles, Mo., have produced videotaped broadcasts on cable television. The videotapes can also be shown to schools and civic groups as part of special public education programs.

Here are some other examples of innovative public education techniques:

- Fire service organizations in Minnesota obtained free space on billboards for fire-related messages.

- The Kingston, Pa., Fire Department printed placemats bearing five different fire service messages and distributed them to local restaurants. One example: "Is your home safe from fire?"

- The department in San Clemente, Calif., enlisted the help of a local women's club to conduct a telephone survey to find out how well the department was reaching citizens with its educational campaign. Then the department mailed out information to people who had not received word on some aspects of fire protection.

- Other departments have developed media kits on specific topics that local broadcast stations can use to inform the public. One developed by the U. S. Fire Administration and the Montana Fire Service Training School dealt with wood-burning stoves and included public service announcements, radio scripts, and 35mm slides.

- The Concord, N.H., Fire Department developed a biweekly column in the local newspaper explaining why firefighters do some of the things they do to suppress fires. The first column, for example, explained how firefighters prevented $5,000 in property damage by chopping a $200 hole in someone's roof.

Public Information

Local news organizations can be important tools in a fire department's public education efforts, but they cannot be controlled as easily as a videotaped fire protection broadcast on the local public service channel or a weekly column the public information officer writes for a local newspaper. Some publicity in the local media might make the department proud, but other stories might detail information chief officers would prefer to forget. That is not to say the news media cannot be influenced, but reporters who are aware of being manipulated tend to become resentful and distrustful, and that is not good for public relations. Cooperation, rather than manipulation, is usually the best tactic.

To have successful media relations, fire administrators should understand how news organizations are structured, and how reporters are trained to work and think. Newspapers, unless they are weeklies or extremely small dailies, will usually assign a specific reporter to cover the fire department. In some cases, two reporters might have different aspects of the department as part of their beats. A city hall reporter, for example, might be responsible for covering fire department administration, while emergency operations might fall into the territory of the police reporter. However the newspaper sets up its coverage, fire department officials and personnel are likely to be dealing with one or two specific reporters or their reliefs on a regular basis. Radio and television coverage of the fire department, on the other hand, is more likely to concentrate on emergency operations than on administration because of broadcasting's smaller staffing levels. The reporters or the camera crews who make a fire run are likely to be the staffers on duty at the time the alarm comes in.

Fire chiefs seldom have to make much effort to get coverage of emergency operations, which reporters will often refer to as "hard" news. Many news organizations will monitor the fire department radio frequency and send reporters and photographers to the incident scene if the call seems important enough to make it newsworthy.

Dealing with reporters on the fireground can be trying. It is hard to direct fire fighting operations while reporters are clustered around asking volleys of questions (Figure 5.7). The fire department has a public duty that should take precedence in all emergencies, but fire administrators should realize that news organizations have a quasi-public function that has been recognized since the ratification of the Bill of Rights almost 200 years ago. Reporters have a right to be on the scene of an incident unless their activity endangers themselves or others or hampers operations.

Figure 5.7 Newspaper and television reporters eager for a news story can make direction of fireground operations more difficult.

One solution is to use public relations to defuse the problem before it crops up in the field. Most reporters are trained, when new to a beat, to visit with sources and find out how things are run. Administrators can bring up fireground problems at this point and suggest ways to alleviate them. If necessary, chief officers can set up more formal meetings with reporters from several organizations to describe what the department does on the fireground and try to come up with mutually acceptable procedures that reporters can use to get their stories on deadline without getting in the way of the firefighters. In their zeal to get stories, reporters often fail to realize the additional problems they cause the department. Be sure to use the meeting to let them know.

The Miami, Florida, Fire Department has devised a novel way to improve its working relationship with the press: it conducts an eight-hour intensive course designed to familiarize media personnel with fire fighting equipment, fireground operations, rescue techniques, and investigative procedures. The results have been positive, and the department has been able to perform some public education.

The department may have to make more of an effort to get coverage of "softer" news, which some reporters will classify as anything that does not offer the excitement of a major fire or a departmental scandal. Reporters nevertheless are on the lookout for feature and human interest stories, and a tip will often be enough to get them started. Newspaper city editors and broadcast news directors appreciate reporters who are able to come up with good stories on their own. Helping them out a little with a tip will make points for the reporters with their supervisors and for the department with the reporters (Figure 5.8). It will also help provide citizens with information they might not otherwise get. If the reporters are not receptive to tips, go directly to the city editor or the news director, the people who pass reporters their assignments.

Reporters are trained to cooperate with their sources, but fire administrators should be aware that reporters are also warned about getting too chummy. They learn to keep the interests of their audience uppermost in their minds, and they are told that sometimes the interests of their audience and the interests of their sources will conflict. The distance is necessary to keep their reporting as unbiased as possible. Journalism schools today are teaching future reporters that they should perform several functions while covering public agencies such as the fire department. One is to inform their audience how to use the services these agencies provide, and this function fits in with the public education goal most departments have. Students are also told they have the responsibility to monitor the performance of public agencies and publicize failures to provide the service the citizens are pay-

Figure 5.8 Be accessible to the news media: provide reporters with general interest information and do not miss a chance to perform some public education.

ing for. Theoretically, such stories are based on hard, verifiable data, not some reporter's imagination. Regardless of the factual basis of the story, news that puts the department in a bad light, deserved or undeserved, is likely to change cooperation into conflict.

Fire administrators have the right to demand accurate reporting, accurate not only in specific details, but in overall impression. They can increase the likelihood a story will be accurate by speaking with reporters in common, everyday language, remembering that reporters are not fire service experts and often do not understand fire service jargon and methods. Accurate stories are also easier to come by if the chief officer or a department representative is accessible to reporters. A "no comment" will not keep a story out of the paper or off the air. Instead, it will send the reporter to other, possibly less-informed, sources for the story, because reporters are taught early that there is more than one way to get information. If, in spite of their efforts, chief officers are still plagued by inaccurate reporting, they should complain vociferously, first to the offending reporter and then, if necessary, to the reporter's supervisor. Such complaints can result in a correction or a retraction. Administrators should keep in mind, however, that reporters expect to have access to public officials and a certain amount of cooperation, even under the most trying and unpleasant circumstances. They resent public officials who are vocal enough when the publicity is good, but who refuse to comment when the story is likely to be uncomplimentary. The department is a public agency, and it is rightfully in the public eye.

Some departments have full- or part-time public information officers (PIO) who are responsible for getting information and distributing it to reporters. On the fireground, the PIO is often the only fire department employee who is authorized to speak with reporters. While many reporters will recognize the necessity of such a rule in an emergency operation, they will resent a PIO who tries to cut off their access to other department officials, including the fire administrator, in nonemergencies. PIOs have a demanding job because reporters expect them to be knowlegeable, informed, and available 24 hours a day.

Some fire department PIOs have been quite successful in working with local news organizations. Los Angeles County had a setup where a public information officer or any other fire captain or battalion chief could radio in a "Code N" when they thought an incident newsworthy enough to contact reporters. Several departments have worked up detailed procedures for contacting news organizations about stories of public interest and what information to release. Departments often will withhold information dealing with personnel matters, details that could hamper an investigation, or medical and biographical information about patients treated by department paramedics under direction of a physician. News organizations should be called in to help define releasable and nonreleasable information.

Fire Department Budgeting

6

NFPA STANDARD 1021
STANDARD FOR FIRE OFFICER
PROFESSIONAL QUALIFICATIONS

Fire Officer VI

7-4.2 The Fire Officer VI, given the pertinent information relative to establishing a fire prevention program, shall:
    (a) establish the goals for a total fire protection program
    (b) determine the budget requirement for implementation of the program
    (c) justify the budget request.

7-4.4 The Fire Officer VI shall demonstrate knowledge of how to determine the funds necessary to operate the department for a fiscal period, prepare a budget as prescribed by the jurisdictional policy, and justify the proposed expenditures to the legislative body.

7-4.5 The Fire Officer VI shall demonstrate knowledge of how to develop and administer a system of budget control based upon fiscal and financial policies of the authority having jurisdiction.*

The above NFPA standards are addressed from a general management perspective.

*Reprinted by permission from NFPA No. 1021, *Standard for Fire Officer Professional Qualifications*. Copyright 1983, National Fire Protection Association, Boston, MA.

Chapter 6
Fire Department Budgeting

The days are long past when fire administrators met with political leaders, asked for a lump of money to run the department for the next fiscal year, and got it. "Budgeting" in those days amounted to taking what the city fathers handed out and trying to make it last until they passed around some more. In these days of accountability, budgeting has to do a lot more than describe what department managers are spending money for and how much more they can spend. Today, the budget and the budgeting process serve as planning and informational tools to help fire administrators and their departments decide the emphasis they want to give certain areas over the next budgeting period. The process also informs city administrators, elected officials, and the public, which provides a good portion of the funding, how the department is being run.

Putting together a preliminary budget for the chief administrator helps chief officers decide the kinds of program improvements and cutbacks they can implement the following year. It also gives them a better grasp on the kinds of long-term changes they need to make (Figure 6.1 on next page). Taking the department's budget into account while hammering together a budget proposal for elected leaders helps high-level administrators decide how the department should fit into the overall job of running the city. Approving the final document in the glare of the public arena forces elected officials to back up policy decisions with hard cash after citizens have had a chance to express their opinions.

TYPES OF BUDGETS

The difference between short-term and long-term planning points up the difference between the two basic components of every budget proposal. Fire administrators have to make deci-

112 CHIEF OFFICER

BUDGET PREPARATION — A FORM FOR PREPARING REVENUE ESTIMATES FOR THE BUDGET

BUDGET 19____   FUND: ____

REVENUES	Actual Collections 6/30/82	Actual Collections 6/30/83	Actual Collections To 12/31/83	Estimated Collections Balance Of Year 1984 July-June	Total Estimated Collections This Year 6/30/84	Estimated Collections New Budget 1984-1985	Collection Estimate Approved By District
TOTALS							

Assumes a Fiscal Year of 6/30

BUDGET PREPARATION — A FORM FOR PREPARING COST ESTIMATES FOR THE BUDGET

BUDGET 19____   FUND: ____

EXPENDITURES	Actual Expenditures 6/30/82	Actual Expenditures 6/30/83	Actual Expenditures To 12/31/83	Estimated Expenditures Balance of Year 1984 July-June	Total Estimated Expenditures This Year 6/30/84	Estimated Expenditures New Budget 1984-1985	Expenditure Estimate Approved by District
TOTALS							

Assumes a Fiscal Year of 4/30

Figure 6.1 Preparing cost estimates for a preliminary budget gives the chief the opportunity to outline long-range goals for the department.

sions about how they will run the department over the next several years.

Details on expected income and planned spending for the next spending year go in a kind of budget that has several names: an expenditures budget, an operational expenditures budget, or an operations budget. Such a budget provides information about ways the chief officer plans to handle recurring purchases, such as uniforms and equipment for firefighters, gasoline and maintenance for departmental vehicles, and clerical supplies for secretarial staff. Budget makers can split an expenditures budget into several accounts to provide the detail managers need to run the department day by day.

The second type of budget, usually called a capital expenditures budget, details the major one-time-only purchases the department plans to make during the coming fiscal year. These are expensive items, such as fire vehicles or even a new fire station, and they often have to be paid for a little bit at a time over several years. Most cities will pay for such items with borrowed money, usually by selling general obligation bonds the voters have approved in an election. The principal and interest on the debt are paid off over 25 or 30 years with annual taxes that go into a sinking fund.

BUDGETING METHODS

Since the municipal reform movement of the early 20th century, administrators have used several methods to account for the public's money.

The Line-Item Budget

Line-item budgeting, probably the most common method of budgeting in municipal administration, lists the department's proposed expenditures line by line in specific and necessarily lengthy detail (Figure 6.2 on next page). Chairs, typewriters, helmets, and hose might all have separate lines in a line-item budget, as well as wages, salaries, and fringe benefits.

Such a method has obvious advantages. For one, administrators can tell at a glance what they have spent and how much they have left to spend. They can total the individual line items to see how much they propose to spend during the next fiscal year, match that figure with the income available, and see if they will be running the department in the red or in the black. Any deficit, of course, can be made up by adjusting property assessments or the tax rate, because the law usually requires cities to operate on a balanced budget, one in which income exactly matches expenditures.

A line-item budget also gives fire administrators a detailed guide for managing the daily operations of their departments.

CITY OF STILLWATER
84/85 BUDGET

DEPARTMENT: 01-730 FIRE SERVICES

ACCOUNT CODE	DESCRIPTION	BUDGETED AMOUNT
1-5-730-101	FULL TIME SALARIES	1,119,178.00
1-5-730-102	PART TIME WAGES	35,752.00
1-5-730-103	OVERTIME PAY	10,000.00
1-5-730-104	ALLOWANCE OR STAND-BY	17,220.00
1-5-730-121	SOCIAL SECURITY	8,608.88
1-5-730-122	RETIREMENT	5,976.00
1-5-730-123	O M L A G INSURANCE	96,498.96
1-5-730-125	WORKMEN'S COMP - PRE	6,527.00
1-5-730-127	UNEMPLOYMENT COMPENS	5,947.50
1-5-730-131	PENSIONS	105,916.60
	CHAR TOTAL	1,411,624.94*
1-5-730-221	MOTOR VEHICLE PARTS	8,000.00
1-5-730-223	BUILDING MATERIALS	1,000.00
1-5-730-234	OTHER EQUIPMENT PART	1,600.00
1-5-730-241	OFFICE SUPPLIES	2,000.00
1-5-730-244	STAMPS & POSTAGE	200.00
1-5-730-245	AGRICULTURAL/HORTICU	150.00
1-5-730-246	JANITORIAL SUPPLIES	3,500.00
1-5-730-247	CONCRETE/SAND/ASPHALT	250.00
1-5-730-249	CHEMICAL	500.00
1-5-730-250	BOOKS & PUBLICATIONS	1,400.00
1-5-730-251	CLOTHING & UNIFORMS	14,000.00
1-5-730-252	FOOD	1,000.00
1-5-730-253	GASOLINE	5,000.00
1-5-730-254	OIL	700.00
1-5-730-255	DIESEL FUEL KEROSENE	1,200.00
1-5-730-258	TOOLS	300.00
1-5-730-259	MINOR SUPPLIES	2,500.00
	CHAR TOTAL	43,300.00*
1-5-730-301	GAS - OKLAHOMA NATUR	3,000.00
1-5-730-304	TELEPHONE	10,000.00
1-5-730-315	OTHER RENTALS	600.00
1-5-730-321	REPAIR OF CARS & TRU	500.00
1-5-730-323	REPAIR OF BUILD. & S	1,500.00
1-5-730-324	REPAIR OF RADIOS	1,500.00
1-5-730-325	REPAIR OF OFFICE EQU	300.00
1-5-730-326	REPAIR OF HEAT & COO	2,500.00
1-5-730-334	OTHER EQUIPMENT	1,000.00
1-5-730-372	DUES & MEMBERSHIP	1,450.00
1-5-730-373	IN-SERVICE TRAINING	4,700.00
1-5-730-374	LAUNDRY	2,500.00
1-5-730-382	MISCELLANEOUS SERVICE	1,000.00
1-5-730-384	TRAVEL	1,200.00
1-5-730-385	PROFESSIONAL SERVICE	5,000.00
	CHAR TOTAL	36,750.00
1-5-730-401	CARS & TRUCKS	11,000.00
1-5-730-402	OFFICE EQUIPMENT/FUR	2,300.00
1-5-730-403	RADIO/COMMUNICATION	10,600.00
1-5-730-405	MAINT/CONSTRUCTION E	6,000.00
1-5-730-407	MAJOR TOOLS	16,150.00
	CHAR TOTAL	46,050.00*
	ACTIVITY TOTAL	1,537,724.94**

Figure 6.2 A line-item budget gives a detailed account of departmental expenditures. *Courtesy of City of Stillwater, Oklahoma.*

The budget tells them what to buy and how much they can spend for it. Chief officers can also use the detail to help put together next year's budget proposal.

Finally, in the plus column, line-item budgets make it almost inevitable that chief officers will spend all the money they have received for the current fiscal year. In cities using the line-item method, administrators frequently operate on the theory that unspent funds were unneeded in the first place, force the offending fire administrator to "turn back" the money that was not spent, and cut the department's budget for the next fiscal year by the appropriate amount. Thus, ensuring that fire administrators spend their appropriations to the last dime is an advantage of line-item budgeting if the goal is to avoid budget cuts the following fiscal year. However, seen from the standpoint of the taxpayers, who more and more are interested in seeing their money spent wisely or not at all, this "advantage" is actually a disadvantage.

This and other disadvantages of line-item budgeting have turned many cities away from the method over the last few decades. In line-item budgets, officials can quickly charge off every expenditure against its appropriate line item, which tells them immediately who the overspenders are. As advantageous as this information may seem from the taxpayer's viewpoint, it can be used by authoritarian managers to clamp down on imaginative and innovative operation. This eventually will hurt the department, the city and, finally, the citizens.

The line-item method also has a problem in focus. It tends to draw attention to individual expenditures rather than to why the money is being spent. The concern is with how many pencils and paper clips the department has to buy instead of what programs the purchases support. Are departments spending their funds to prevent fires or to put them out? Do they spend money for program improvements that eventually will meet department goals? In other words, a line-item budget gives little information about purpose. The citizens and political leaders who examine such budgets become more interested in finding waste and cutting it out than in realistically questioning why the department wants to spend the money it is asking for. Seldom do they ask if there are other ways to arrive at the same end with less expense. As a result, departments get locked in to today's methods, and the way the department does it today becomes the way it does it tomorrow. If no one can figure out why departments are spending money, it is difficult to find out if they are spending it wisely.

The Performance Budget

In this kind of budget, fire administrators usually split the department into different services or functions: fire prevention, suppression, training, and others as necessary. Then they have to come up with performance standards and compare them with what their departments actually do to see how they measure up (Figure 6.3). For instance, chief officers might decide that the department, considering the proposed staffing and appropriations, should be able to conduct 120 fire inspections during the next fiscal year and this would be an acceptable level of operation. Then the chief officers divide the number of proposed inspections into the proposed cost to come up with a "unit cost." The idea behind all the figuring is to cut the unit cost, maintain the status quo or at least limit the unit cost increases from year to year.

The advantage of this budgeting method is that it shifts the focus from individual purchases to the reason why departments made the purchases and the results they expect to get. Departments decide what kinds of improvements to make during the funding period and then, as the fiscal year winds down, try to evaluate what they accomplished. Of course, with the performance standards at hand policy makers and the voters can also do a better job of evaluation by putting more pressure on the fire administrator to justify how the appropriation was spent.

This budgeting method is also loaded with problems, which are generally considered to be pretty serious, and there have been few changeovers to program budgets since the late 1950s.

The first question concerns who establishes the performance standards. The tendency has been for authoritarian managers to come up with the standards without asking the opinion of too many of the department employees. This is bad for morale, and eventually can be inaccurate and inefficient.

Once the standard makers are identified, the next problem becomes quantifying operations that may be hard to express in numbers or that may be meaningless if they are. For example, what does it mean to say that firefighters put out more fires this year than last? Is this an increase in efficient fire suppression or a decrease in efficient fire prevention?

When officials come up with the types of work units they want to count, they usually have to hire more employees to count the work units, and the cost of extra personnel will often eat up any savings resulting from the necessary large-scale review of what the department wants to do and actually does.

Finally, the performance budget gives little attention to the question of where the fire department fits into the overall operation of the city. Also ignored is how one city department works in cooperation with, or perhaps at counterpurposes to, another.

The Program Budget
The program budget, which has developed since the end of World War II, gets rid of all or most of the line items that used to

Fire Department Budgeting

PERFORMANCE INFORMATION TRACKING SYSTEM QUARTERLY REPORT

DEPARTMENT:	FIRE				PROGRAM #2305XX
PROGRAM:	FIRE SAFETY/TRAINING				

Workload Measures		Cumulative Total 1982-83	1983-84 Q1	1983-84 Q2	1983-84 Q3	Cumulative Total 1983-84
1. Training sessions per company	Target	120	30	60	90	120
	Actual	120	30	60	90	120
	% Attainment	100%	100%	100%	100%	100%
2. No. of specialized courses	Target	15	3	7	11	15
	Actual	20	2	6	9	16
	% Attainment	133%	67%	86%	81%	107%
3. No. of recruit training hours	Target	9,308	0	6,720	0	6,720
	Actual	0	0	6,720	0	6,720
	% Attainment	0	—	100%	—	100%
4. No. of new or upgraded manuals	Target	3	0	1	2	3
	Actual	6	0	1	1	3
	% Attainment	200%	—	100%	50%	100%
5. No. of personnel evaluated	Target	1,200	300	600	900	1,200
	Actual	1,400	300	600	900	1,200
	% Attainment	117%	100%	100%	100%	100%
6. No. of apparatus tested	Target	45	0	0	0	45
	Actual	5	40	0	0	64
	% Attainment	11%	—	—	—	140%
7. No. of equipment test on equip.	Target	135	30	70	110	135
	Actual	135	30	70	110	135
	% Attainment	100%	100%	100%	100%	100%
8. No. of tests used to investigate new equipment or procedures	Target	25	8	12	20	25
	Actual	25	6	10	15	30
	% Attainment	100%	75%	83%	75%	120%
Effectiveness Measures						
1. Hours lost due to injury leave	Target	—	—	—	—	—
	Actual	13,968	5,114	5,928	3,864	20,090
	% Attainment	—	—	—	—	—
Productivity Measures						
1. No. of hours lost — no. of employees	Target	—	—	—	—	—
	Actual	6.70	10.95	12.03	8.45	10.71
	% Attainment	—	—	—	—	—

Figure 6.3 The performance budget allows comparison of different departmental expenditures. *Courtesy of Long Beach, California Fire Department.*

clutter up budgets and instead appropriates all funds within a particular program for a few broad purposes (Figure 6.4). Typical are personal services, which include wages, salaries, and fringe benefits; capital expenditures to purchase costly items; interfund transfers, a bookkeeping device that allows department managers to transfer funds from one program to another if necessary; and services and supplies. Part of the purpose is to turn attention away from the nuts-and-bolts detail and toward purpose and the levels of service the public, the politicians, and city officials can live with and afford.

This method, like the performance method, tends to emphasize programs instead of how many reams of typing paper the department plans to purchase during the fiscal year. Fire administrators can use the evaluation this approach requires to identify duplications and reorganize to eliminate them. Finally, the program approach forces officials to set priorities for the department by deciding how much they want to spend on each program.

The disadvantages of the program budget eventually forced modifications, which will be described below. For one thing, poor program definition will hamper the method's usefulness. It is also hard to decide on priorities without some way to measure work load; if this kind of detail is added, the more control-oriented types in the budget office can put the same restriction on creativity described earlier in the section on line-item budgets. Finally, the method lacks the detail fire administrators need to operate their departments from day to day; however, if the detail is added, chief officers can once again find themselves in conflict with their budget offices.

Another form of the program budget seems to be a merger of the type discussed above with the line-item budget. This form of the program budget gets away from the broad categories of spending, such as personal services and capital expenditures, and provides more line-item detail under each department program.

One advantage is that fire administrators can see at a glance all the costs of a specific program and thus can get an idea of what effects program changes will have on expenditures. If something unexpected comes up, they can more easily decide which programs to cut. Another advantage is that this method forces administrators to come up with a uniform way of determining actual costs throughout all programs in a department and throughout all departments in the city.

One problem is coming up with a system to distribute costs to various programs. If officials are willing to take the time and effort, they can come up with programs to handle the details by computer. Another, more fundamental problem, is that this program budget variation tends to take control of the budget away

FIRE DEPARTMENT PROJECT BUDGET — 1984

PROGRAM COMMUNITY FIRE EDUCATION

PROGRAM DESCRIPTION Educating community about fire hazards and heightening awareness of fire safety.

PERFORMANCE OBJECTIVES

1. Educate public about basic fire safety issues.
2. Conduct presentations on fire safety to schools, church groups, civic groups.
3. Conduct private home inspections and hold discussions on residential smoke detectors.
4. Conduct CPR classes.

INDICATORS OF PERFORMANCE

MEASUREMENT	OBJECTIVE	1984 ACTUAL	1985 ESTIMATE	1986 PROJECTED
DEMAND				
Home Inspections	75	125	150	175
Civic Groups — Presentations	15	16	19	20
School Demonstrations and Presentations	10	10	10	10
WORKLOAD				
Hours — Firefighter	300	420	450	500
Hours — EMT	75	100	125	150
Hours — Officers/Inspectors	200	275	300	350
PRODUCTIVITY				
Requests for home inspections	50	37	50	50
Requests for tot finder stickers	75	41	75	100
Requests for information on smoke detectors	500	300	650	700
EFFECTIVENESS				
Fire incidents — children and matches	10	4	6	3
Number of faulty smoke detectors	30	32	25	20
Chimney fires	10	5	4	3

ANALYSIS

The City of Lathem Fire Department, Division of Fire Education will implement this program with the assistance of various fire companies and EMS units. This program is a budgeted item of the 1984 Fire Department Annual Budget.

The goal is to educate the community about common fire hazards within the home and to point out the resources that are available from the city.

Effectiveness will be measured by actual numbers of fire incidents that are recorded for different classifications. In addition, a measure will be performed of public requests for materials, presentations, inspections, equipment demonstrations, and films.

RESOURCES

CATEGORY	1984 ACTUAL	1985 BUDGET	1985 REVIEWED	1986 BUDGET
Personnel ($)	1100	1500	1750	1750
Film Rental/Purchases	100	250	350	400
Printed Materials	500	700	750	800
Equipment/Supplies	255	450	500	500
Refreshments	495	500	600	700
TOTAL	**2450**	**3400**	**3950**	**4150**

Figure 6.4 In a program budget, funds are assigned to several broad areas within a particular program.

from the budget office and distribute it to people in charge of programs. Some jurisdictions are not flexible enough to handle such a dispersion of power. Finally, this form requires even more record keeping than a line-item budget because of all the cost prorating it entails.

The Planning-Programming-Budgeting System

This version of the program budget, introduced by the federal government and then essentially dropped in 1972 after a decade's use, also required managers to define programs within a department and set objectives for each one. Then officials had to detail all the possible ways the department could meet objectives and the costs for each alternative. They compared the costs of each alternative to the benefits and chose for funding the best plan for reaching the objectives, and managers programmed each year of the plan into that fiscal year's budget.

One major advantage of this system, known as PPBS, was that it forced officials to make a systematic, detailed study of the cost-benefit ratios of different ways to meet the same goals. With PPBS, officials had to consider alternatives. Spreading the plan to meet the objectives over several years theoretically ensured operational continuity, something that had been lacking in some other budgeting methods.

PPBS seemed ideal in theory, but there turned out to be some practical limitations. Some of the programs were so poorly defined in practice, that they were overseen by not only the program analysts, but by more than one manager. There were few line-item details for daily operations. There were not enough trained analysts to do the job. The method generated mounds of cost-benefit analyses. It was practically a physical and mental impossibility to come up with every conceivable alternative for anything and everything a city or a fire department wanted to do.

These were not serious problems compared to this last one: PPBS was based on the idealistic assumption that political leaders would choose the program alternative with the best cost-benefit ratio. In the real world of politics, constituencies, and special interest groups, favorable cost-benefit ratios were among the last things elected officials considered.

The Integrative Budgeting System (IBS)

Budget makers eventually came up with this pick-and-choose system in which they took all the workable elements of other budgeting methods and ignored the rest. This system, for example, breaks up a department into several functions or programs, but categorizes each program only into personal services, maintenance and operation, and capital expenditures accounts (Figure 6.5). There is some line-item information, but it is there

LINE ITEM BUDGET AND PROJECTIONS

DEPARTMENT: FIRE **DESCRIPTION:** FIRE SUPPRESSION **FY 1985**

	ACTUAL EXPENDITURES	APPROPRIATION	CODE	OBJECT DESCRIPTION	RECOMMENDED FY 85-86	+(-)	APPROVED FY 85-86	PROJECTIONS FY 86-87	FY 87-88	FY 88-89	FY 89-90
FY 82-83	FY 83-84	FY 84-85									
				PERSONAL SERVICES							
85	87	87	10.000	Full-Time Positions	87			87	87	87	87
848	712	712	11.000	Regular Overtime Hours	640	-62		640	640	640	640
848	712	712		TOTAL HOURS REQUIRED	640	-62		640	640	640	640
				PERSONNEL COSTS							
1,464,677	1,529,113	1,701,881	10.000	Salaries	1,731,484	29,603		1,818,058	1,908,960	2,004,408	2,104,628
105	956	39,580	10.001	Supplemental Pay	43,990	4,410		46,189	48,498	50,922	53,468
1,228	316	10,372	10.010	Regular Overtime	9,339	-1,033		9,805	10,295	10,809	11,349
6,365	8,331	3,022	10.015	Paramedic Overtime	3,022			3,173	3,331	3,497	3,671
1,783		15,010	10.100	Constant Manning	10,010	-5,000		10,510	11,035	11,586	12,165
			10.150	Part Time							
59,226	61,302	66,685	10.200	Special Pay	75,010	8,325		78,760	82,698	86,832	91,173
329,711	345,015	399,356	10.300	Retirement	406,232	6,876		426,543	447,870	470,263	493,776
	219	384	10.350	Supplemental Retirement	414	30		434	455	477	500
9,330	27,224		11.000	Injury Pay	20,853	-6,371		21,895	22,989	24,138	25,344
40,701	45,603	52,091	11.100	Health Insurance	70,594	18,503		74,123	77,829	81,720	85,806
10,897	10,931	11,485	11.200	Life Insurance	12,260	775		12,873	13,516	14,191	14,900
66,376	42,447	108,865	11.300	Workers Compensation	59,560	-49,305		62,538	65,664	68,947	72,394
7,654	7,725	10,306	11.400	Dental Insurance	10,306			10,821	11,362	11,930	12,526
17,704	24,197	26,193	11.050	Sick Leave Reserve	26,445	252		27,767	29,155	30,612	32,142
			12.000	Vacation Reserve	13,227	13,217		13,888	14,582	15,311	16,076
2,006,427	2,085,485	2,472,454		**TOTAL PERSONNEL SERVICES**	2,492,746	20,282		2,617,377	2,748,239	2,885,643	3,029,918
				MAINTENANCE AND OPERATION							
13,614	18,831	21,850	20.100	Utilities	24,010	2,160		25,210	26,470	27,793	29,182
51,746	51,746	53,628	22.000	Hydrant Rental	9,128	-44,500		9,584	10,063	10,566	11,094
1,348	1,870	1,848	24.000	Office Supplies	1,940	92		2,037	2,138	2,244	2,356
943	951	970	24.500	Office Equipment Expense	970			1,018	1,068	1,121	1,177
12,530	10,905	12,960	25.000	Uniforms	13,510	550		14,185	14,894	15,638	16,492
74,650	75,344	87,131	26.000	Equipment Expense	100,075	12,944		105,000	110,000	115,373	121,026
69	448	483	26.050	Small Tools	383	-100		402	422	443	465
117	425	388	28.000	Travel & Subsistence	388			407	427	448	470
340	380	430	29.000	Training	430			1,060	1,113	1,168	1,226
	1,010	1,010	29.100	Firefighters Olympics	1,010			313	328	344	361
114	71	299	30.000	Dues & Subscriptions	299			509	784	950	1,040
8,593	18,917	13,660	40.000	Special Expense	13,660			2,110	2,215	2,325	2,441
597	809	2,010	10.700	Paramedic Program	2,010			14,343	15,060	15,813	16,603
199,728	219,351	296,252	30.000	Overhead	295,301	-951		310,066	325,569	341,847	358,939
364,389	401,058	492,919		**TOTAL MAINTENANCE AND OPERATIONS**	463,114	-29,805		486,244	510,551	536,073	562,872
				CAPITAL OUTLAY AND IMPROVEMENTS							
			40.000	Fire Hose	8,010	8,010					
			50.000	Breathing Apparatus	1,630	1,630					
			27.000	Firefighting Tools	850	850					
20,446	28,599	15,830	49.999	Prior Year/Projections		-15,830					
20,446	28,599	15,830		**TOTAL CAPITAL OUTLAY AND IMPROVEMENTS**	10,490	-5,340					
3,392,110	2,515,854	2,981,915		**TOTAL APPROPRIATIONS**	2,966,990	-14,925		3,104,261	5,259,430	3,422,356	3,593,430

Figure 6.5 Integrative budgeting systems combine elements of all budgeting systems to analyze departmental needs on one large scale.

only to help managers with daily operations of their departments, not to help elected officials make policy or high-level administrators set up service levels. The line-by-line cost estimates are used to come up with a total figure for the program that administrators ask elected officials to appropriate without tying any amounts to specific items. After officials define the programs and the expenses tied to them, they spread overhead and support costs over all the programs in the department. Instead of performance standards, managers use performance "indicators" to see how they are meeting department objectives. Budget makers retain the formulas for computing cost-benefit ratios but apply them to specific programs only if studies indicate the benefits in increased productivity would be worth the time, effort, and money required to conduct the analysis.

IBS has the same advantages as the other systems from which its component features are drawn. In addition, it is computerized (it has to be), and it uses behavioral controls instead of authoritarian controls, which management theorists consider to be more productive.

Such a system is also difficult and time-consuming to implement. It can take up to three or four years to set up an IBS and put it into operation. Because it requires a complete overhaul of the budgeting system then in effect and the initiation of behavioral controls it can be threatening to the existing power and control structure of the organization. The system also requires policy makers to decide what service levels they want, something vote-conscious politicians are often reluctant to do.

Zero-Based Budgeting (ZBB)

Introduced in the federal government in the late 1970s, ZBB is one of the latest budgeting methods. It has been described as a start-over approach assuming there is no such thing as a fire department once a year at budget time and building one up from the ground. This is an oversimplification. What ZBB actually involves is a new set of terms for old concepts and four different budget proposals for four alternative service levels (Figure 6.6).

The approach starts with program definition, although now the programs are called "decision units." Managers compile the decision units into what is called a "decision package," formerly called a department budget, that includes the following:

- A statement of what the program is designed to do
- A description of how officials hope the package will fulfill its goals
- An estimate of how much it will cost to do what the officials hope to do and the benefits expected to result
- A method of evaluating work load and performance

CITY OF PHOENIX, ARIZONA
MANAGEMENT AND BUDGET DEPARTMENT
BUDGET DECISION PACKAGE
1984-85 FISCAL YEAR

PROG	FUND	DECISION PACKAGE NAME	DIST	LEVEL	B/S	POS	NET COST	DEPARTMENT RANK	MGR RATING
FI	01	Division Chief — Haz Mat	0	1/1	B	1.0	$55,150	13 of 62	

DIVISION	DIVISION RANK	SECTION
Emergency Services	1/5	Hazardous Materials

DESCRIPTION

Division Chief — Hazardous Materials Program Coordinator

COST DATA

PERSONAL SERVICES	55,150
CONTRACTUAL SERVICES	
COMMODITIES	
CAPITAL OUTLAY	
GROSS COST	55,150
EXPENDITURE CREDITS	
REVENUE CREDITS	
NET COST	55,150
RECURRING ANNUAL COST	55,150

PROGRAM IMPACT — BENEFITS OF FUNDING AND CONSEQUENCES OF NOT FUNDING

Benefits: Funding would maintain current level of Program Approach to Management of Hazardous Materials Response and Intervention Strategies. This includes development of Haz Mat Training, Haz Mat Incident Scene Command, Inter-Agency Coordination, and Management and Analysis of Data.

Consequences: Not funding would result in loss of Section Manager for Haz Mat Program. This would severely reduce the ability of the Department to successfully mitigate Hazardous Materials Emergencies.

POSITION DATA

FULL-TIME	PART-TIME (FTE)
1.0	

HIRING DATE

PROG	FUND	DECISION PACKAGE NAME	DIST	LEVEL	B/S	POS	NET COST	DEPARTMENT RANK	MGR RATING
FI	01	Command Van Engineers	0	1/1	B	3.0	$102,054	14 of 62	

DIVISION	DIVISION RANK	SECTION
Emergency Services	2/5	Firefighting

DESCRIPTION

Engineer (3) Command Van Driver

COST DATA

PERSONAL SERVICES	102,054
CONTRACTUAL SERVICES	
COMMODITIES	
CAPITAL OUTLAY	
GROSS COST	102,054
EXPENDITURE CREDITS	
REVENUE CREDITS	
NET COST	102,054
RECURRING ANNUAL COST	102,054

PROGRAM IMPACT — BENEFITS OF FUNDING AND CONSEQUENCES OF NOT FUNDING

Severely reduces command function at emergency incidents by restricting incident commander's capability to manage information and resources (manpower and equipment).

This position is extremely vital during complex operations.

POSITION DATA

FULL-TIME	PART-TIME (FTE)
3.0	

HIRING DATE

Figure 6.6 In zero-based budgeting, the department account is set at zero and then four contingency plans are calculated for the budget. *Courtesy of City of Phoenix, Arizona.*

- A list of other methods to accomplish the same goals
- A prediction of what will happen if the package operates at various levels of effort

Managers rank each decision package, each collection of decision units or programs. If policy makers approve a decision package for funding, officials supply it with a detailed operating budget. The policy maker's decision is based on the four budget levels city staffers provide with each decision package: current, which includes budget hikes for expected inflation and known cost increases; minimum, the least amount the department can operate on, usually 65 percent; reduced, about 80 or 90 percent of current; and improved, about 110 percent or more of current.

With ZBB, theoretically, elected leaders and their constituents will know exactly what they are getting for their dollars. They will be able to see the effects of reduced or increased expenditures in terms of reduced or increased service levels. They will no longer be able to criticize the fire administrator for not providing services the taxpayers have refused to pay for, a perennial complaint. Officials will also be able to see exactly what extra dollars will buy and decide whether the benefits are worth the costs. If annexation requires program expansion or an emergency forces fund cuts, administrators will have a good idea of what will happen to service levels. There are also advantages in the program review and rank-ordering the method requires, and it puts a premium on coming up with new methods to meet old objectives.

There are disadvantages, too. Ranking programs is a matter of opinion, and opinion can differ from one level of administration to the next. Another problem is defining decision units properly, making sure that some of the program's costs are not hidden in other parts of the budget. All too often in practice, administrators have used ZBB only to justify existing methods of doing things and their failure to come up with new, creative ways to operate. ZBB also takes trained people and a great deal of time to set up. Some have estimated that cities need three years to set up the system to see if it will work in their specific situation.

THE BUDGETARY PROCESS

Fire administrators at budget-setting time compete with other department heads for increasingly scarce funds. That makes the budgetary process, almost by definition, a political one. While fire administrators might not have to indulge in Machiavellian tactics to get their rightful share of the budget pie, they will find themselves required more and more to justify their budget requests with hard, unemotional facts (Figure 6.7).

When putting together a preliminary budget for the upcoming fiscal year, preferably with input from other fire department personnel, fire administrators will often start with budgets from

Figure 6.7 Fire protection is one of the many services a city must provide, so budget requests must be backed by facts and careful planning.

earlier years, checking to see how accurate they were and how relevant old expenditures are for the current situation. Unneeded expenses can be cut, and new needs written in. Budget makers must also consider increases in the cost of living and inflation, and whether the department needs to make any capital purchases.

The preliminary budget goes to the chief administrative or executive officer of the city, be that the mayor or a professional city manager, or to one or more of several assistants. These assistants may be called budget officers, budget managers, finance officers or assistant city managers for finance. The preliminary budget could also go to a finance committee. This early proposal might go back and forth between the department and the administration several times, with several compromise versions, before all preliminary budgets are put into a proposed budget package for elected city officials to consider. Department heads might be called on to explain and justify their proposals in public budget sessions before the elected officials adopt the budget and, in some jurisdictions, adjust the tax rate to fund it if necessary.

Describing the process is simple, but getting what the department needs in a time of fiscal restraint takes savvy. David B. Gratz has listed several methods fire administrators can use to justify their preliminary budgets:

- Instead of harping on the costs, describe the program and its expected benefits. Once administrative and elected officials accept the program, they are more likely to come up with the money to fund it.

- Try to identify fixed, recurring expenses. Once officials are convinced the expenditures will come up regularly, they will not be in controversy when the department budget comes up for approval in following years. Such fixed expenses include fringe benefits packages, retirement fund contributions, depreciation and systematic replacement of equipment.

- Let officials know how spending funds in a certain program now will bring savings in the future.

- Use what has happened in the past to justify more spending in a problem area. Be prepared to back up all such requests with hard, reliable data from the department's record-keeping system.

- Compare the department's operations and situation with those of other departments in nearby cities or adjoining regions. Be sure to use this approach with other tactics because city officials are usually resistant to this line of reasoning by itself.

- Miss no opportunity to impress city officials with your managerial abilities. Being able to operate efficiently with less and demonstrating a sincere desire to cut costs without cutting services can go a long way toward making the department's preliminary budget request more palatable.

There are other methods that department heads have used to justify their budget requests, but Gratz advises fire administrators not to use them because of their questionable professionalism. Here are the ones he lists:

- Sticking unneeded items in the budget so administrators and elected officials can find them and, with much public self-righteousness, cut them out. Behind this tactic is the assumption that the preliminary budget will be cut regardless so the fire administrator might as well inflate it and get the department what it needs. Fire administrators who can justify their requests can often get what they want if the city has no higher priorities.

- Inflating costs in one program so the chief officer can later transfer the excess to another program that had less of a chance of being approved at the desired level.

- Using the emotional parade-of-horribles technique claiming that if administrators fail to give the chief officer what the department needs, the city faces another Cocoanut

Grove disaster, the 1942 Boston nightclub fire in which 492 persons died.

BUDGET CONTROLS

Once the department's budget is adopted, the fire administrator can expect the budget office to monitor department expenditures with a variety of controls that can be classified as traditional or behavioral. The traditional types are used, and sometimes abused, most often. The behavioral types are considered more effective.

Traditional budget controls include the following techniques:

- Line-item accounting, where every expenditure is deducted from the amount remaining in the account. This method usually puts a burden on department managers, because while the managers get blamed for any overspending, the controls on spending usually are vested in the budget office.

- Budgetary accounting reports, usually monthly or quarterly, showing how much has been spent for what and how much is left in each account.

- Percentage deviation reports that show the percentage of allocations already spent compared to what "normally" should have been spent during the time period. This control operates on the erroneous assumption that if the fiscal year is, for example, one-third over, no more than one-third of the allocation should have been spent.

- Allotments that split up the allocation into quarterly or monthly segments and allow department heads to spend no more than that during the allotment period without tussling with the budget office.

- Position controls, usually required because most city departments spend a majority, up to 95 percent, of their funds on personal services. This control limits the number of positions, thus the number of employees, thus the amount of money that can be spent on this major budget item.

- Purchase order and contract award review that requires department heads to get approval from a purchasing office and perhaps from a central budget office before buying budget-authorized items. With this system there is a possibility that purchases approved when the budget was adopted can be turned down when the fire administrator tries to spend the money.

- Travel and subsistence clampdowns.

- Training, dues, and subscription limitations.

Among behavioral controls:

- Motivation often provided by letting more people have a say in setting priorities and seeing that they are met.
- Management by objectives that helps people understand why they do what they do and generally causes them to work more efficiently and cooperatively, especially since they have had a hand in setting up the goals and objectives of the organization.

ALTERNATIVE SOURCES OF FUNDS

In the past, fire administrators have not had to worry about generating funds for their department's operations. Taxpayers expect to get fire protection out of the taxes they pay without having to cough up an extra fee before firefighters will put out a fire at their home. Traditionally, citizens have paid property taxes to fund their city's operations, including the fire service. Some cities, in recent years, have had to levy income taxes to make up the difference between income and the rising costs of services. In the 1970s, as taxpayers became more and more outraged at the bite government was taking out of their paychecks, tax funds became harder to come by; cities have had to dream up other ways to get money.

The trend is expected to continue into the waning years of the 20th century. In the early 1980s more states passed legislation that limited or reduced the taxes on which municipalities have normally relied. California, which initiated the so-called taxpayer revolt of the '80s with the passage of the Proposition 13 limitation on property taxes, continued the trend by passing a limit on how much money local governments can appropriate in a fiscal year (allowing for population and cost-of-living changes). Iowa, Michigan, Oregon, and New Jersey also approved tax limitations of various forms; and Georgia, Nevada, South Dakota, and Virginia approved tax reductions for certain classes of citizens, such as the elderly.

William A. Ward listed four forces contributing to shrinking revenues and budget crunches as America moved into the 1980s:

- Money is not coming in from taxpayers like it used to. Economic conditions make it impossible for taxpayers to pay higher taxes and still keep up an acceptable standard of living.
- The cost of government is going up because the cost of providing services has been going up.
- The demand for services has stayed the same or increased.

- The taxpayers in increasing numbers are blaming government and its cost for more and more of their financial woes. For angry, unhappy, combative taxpayers, local government is an accessible point of attack, much easier than government at other levels.

As local government moved into a period of cutback and retrenchment, fire departments too have had to come up with ways to generate some of their own income as heavier demands are placed on general fund revenues.

Service Fees

Some have suggested setting up schedules of fees for different services fire departments perform. One suggestion from California would base the schedule on the amount of water used to put out a fire. The proposal was designed to encourage property owners to install fire warning devices, such as smoke detectors, and fire control devices, such as automatic sprinkler systems, by establishing lower rates for owners who accepted the expense of installing them.

Others have suggested combining the city's building inspection division with the department's fire prevention bureau. The proposal would require annual inspections of occupied buildings. Fees would be used to support fire department operations (Figure 6.8 on next page). Other portions of the proposal would permit the department to levy the costs of putting out deliberately set fires against those who set them, issue and charge for permits to transport hazardous materials inside city limits, and assess developers for business activity that would increase the department's operational costs.

One group went so far as to have a bill introduced in the Idaho legislature that would allow fire departments to charge fees for fighting fires at tax-exempt institutions, such as churches and hospitals. Although church leaders effectively killed the bill by getting legislators to table the measure, it was one of the few service fee propositions that got beyond the talking stage. A fee schedule developed by Boston was struck down by the state supreme court on the grounds that it constituted double taxation.

Public and Private Grants

There are some public and private funds available for innovative fire service projects, but as the country moved into a period of fiscal conservatism in the early 1980s, the public sources, at least, began to dry up. The federal government began cutting back on local government spending in 1979, and allocation of general revenue sharing funds began to slow down. While there was an increase in general revenue sharing in 1979 and 1980 over what was spent in 1978, it was the smallest increase in

FEES FOR ISSUANCE OF PERMITS

U.F.C. Section	Type of Permit	Fee	Duration
10.307	Fire Alarm Systems	$ .40	Each Head
		25.00	Minimum
10.308	Fire Extinguishing Systems	25.00	Each Occurrence
10.309	Automatic Sprinkler Systems	.40	Each Head
		25.00	Minimum
10.315	Exhaust Hood and Duct Systems	25.00	Each Occurrence
11.101	Incinerators and Open Burning	10.00	Each Occurrence
24.102	Airports, Heliports and Helistops	40.00	Annually
25.101	Places of Assembly		
	50-300 persons	30.00	Annually
	Over 300 persons	50.00	Annually
26.102	Bowling Alleys	25.00	Annually
27.102	Cellulose Nitrate Plastic (Pyroxlin)	25.00	Until Revoked
28.102	Storage and Handling Combustible Fibers	25.00	Until Revoked
29.102	Repair Garages	25.00	Annually
30.101	Lumber Yards	25.00	Annually
30.105	Woodworking Plants	25.00	Annually
31.102	Tire Rebuilding Plants	25.00	Annually
32.101	Tents and Supported Structures	25.00	Each Occurrence
33.102	Cellulose Nitrate Motion Picture Film	25.00	Until Revoked
34.102	Automobile Wrecking Yards, Junk or Waste Material Handling Plants	25.00	Annually
45.102	Flammable Finishes	25.00	Annually
46.102	Fruit Ripening Processes	25.00	Annually
47.102	Fumigation and Thermal Insecticidal Fogging	30.00	Each Occurrence
48.102	Magnesium	25.00	Annually
49.101	Welding and Cutting	25.00	Until Revoked
50.103	Organic Coatings	25.00	Annually
62.102	Ovens, Industrial Baking and Drying	25.00	Until Revoked
63.103	Mechanical Refrigeration	25.00	Annually
74.103	Compressed Gases		
	0-2,0000 Cu. Ft.	25.00	Until Revoked
	Over 2,000 Cu. Ft.	25.00	Annually
75.103	Cryogenic Fluids	25.00	Until Revoked
76.102	Dust Producing Operations	25.00	Annually
77.104	Explosives and Blasting Agents	25.00	Annually
78.102	Fireworks Displays	50.00	Each Occurrence
	Over 6-inch Aerial Bust	75.00	Each Occurrence
79.103	Flammable or Combustible Liquids	25.00	Until Revoked
79.113	Removal of Underground Tanks		
	0-5,000 Gal.	15.00	Each Tank
	5,000-10,000 Gal.	20.00	Each Tank
	Over 10,000 Gal.	25.00	Each Tank
79.601	Installation of Underground Tanks		
	0-5,000 Gal.	20.00	Each Tank
	5,000-10,000 Gal.	25.00	Each Tank
	Over 10,000 Gal.	30.00	Each Tank
79.901	Service Stations	25.00	Annually
79.1201	Tank Vehicles	25.00	Annually
79.1401	Bulk Plants	40.00	Annually
79.1501	Chemical and Processing Plants	50.00	Annually
79.1601	Refineries and Distilleries	50.00	Annually
79.1701	Transportation Pipelines	25.00	Until Revoked
79.1801	Dry Cleaning Plants	25.00	Annually
80.102	Hazardous Materials	25.00	Annually
81.103	High Piled Combustible Stock	20.00	Annually
82.102	Liquefied Petroleum Gases	25.00	Until Revoked
82.102	Liquefied Petroleum Plants	50.00	Annually
83.101	Matches	25.00	Annually
84.101	Motion Picture Projection	25.00	Until Revoked

INSTITUTIONAL OCCUPANCY FEES

More than 50 persons except day care nurseries	$75.00	Annually
More than 6, but less than 50, persons except day care nurseries	50.00	Annually
Under 7 except day care nurseries	25.00	Annually
All day care nurseries more than 6	25.00	Annually

FIRE FLOW TESTS

Fire Flow Tests for Automatic Sprinkler Systems	$50.00	Each Occurrence

Figure 6.8 Some localities are raising revenues by charging fees for performing various types of inspections and permits. *Courtesy of Napa, California Fire Department.*

years. Considering the value of 1980 dollars compared with 1972 constant dollars, increased expenditures actually amounted to a decrease in buying power. The trend was expected to continue through the 1980s.

One group that tried to tap private funds for the fire service was the Idaho Conference of Fire Prevention Officials. It became a nonprofit organization in 1978 to solicit private funds for fire prevention activities after the legislature failed to fund the state fire marshal's office.

Self-Insurance

Under this alternative to general fund support, a fire department would operate just like a fire insurance company, collecting premiums for policy holders, paying off on claims, and supplying fire protection systems that would be installed in policyholders' structures. Proponents claim the detection systems should help cut losses, and the premiums could eventually make the fire department self-supporting. Few, if any cities have tried to get into the fire insurance business, which would probably precipitate a court challenge from the insurance industry, but a survey conducted studies to find out what the plan might do if adopted. One actuarial study conducted by Mountain View, Calif., then a city of 60,000, indicated that within seven years the city would generate enough income from premiums to

- Have smoke detectors in all residences
- Set up a $2.7 million major catastrophe fund that would earn about $161,000 a year
- Have enough excess premiums to fund 65 percent of the city's fire protection service

Regardless of what funding options are available, it seems certain that for the next few years at least part of the fire administrator's budgetary duties will include coming up with new sources of income for fire department operations.

Emergency Medical Services

7

CHAPTER 7 OBJECTIVES

1. Evaluate the EMS program for effective and efficient coordination between the fire department, ambulance services, and hospital emergency personnel.

2. Coordinate necessary standard operating procedures and protocols according to the level of EMS provided in the area.

3. Coordinate work schedules to maintain identified level of EMS.

4. Analyze EMS run statistics to identify potential service or personnel problems and to project future staffing needs.

5. Analyze community EMS needs and project costs and benefits for providing upgraded levels of EMS.

6. Coordinate continuing education programs designed to maintain skill levels of EMS personnel.

7. Develop public education programs to educate community about EMS services provided and to reduce system abuse.

Chapter 7
Emergency Medical Services

EMS AND THE FIRE SERVICE

The report of a heart attack, an automobile accident, or any other medical emergency brings the same competent life safety response as the report of a fire. During the past decade, the fire service has become increasingly involved in emergency medicine (Figure 7.1). In spite of some initial resistance, fire departments

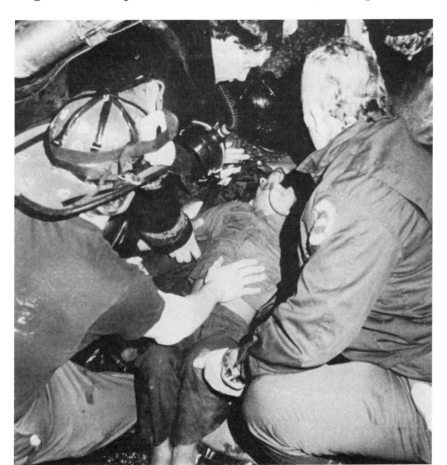

Figure 7.1 Firefighters administer oxygen to a victim of smoke inhalation.

are finding that operating their own emergency medical service is valuable. It has not only saved lives, but also has helped gain public support and improved the image of the fire service. Many fire department administrators have also found it a strong selling point in staffing requests, budget hearings, and other forms of revenue considerations.

Fire departments started providing emergency medical services years ago when the general public began seeking fire department assistance after seeing the results of the resuscitators used to aid firefighters suffering from smoke inhalation. The role of the fire service in emergency medical services has changed greatly since then, with emergency medicine emerging as a specialty and many fire departments requiring all members to be certified as EMTs.

The individual roles that fire departments have in the EMS system and their methods of fulfilling those roles vary depending on the area served, departmental policies, local and state laws, and other factors. In areas where an ambulance may have extended response time, the fire department may send an engine company as a first responder to provide care for the victim until the ambulance arrives. A department in a smaller response area may send its own ambulance or rescue apparatus and personnel trained to provide advanced life support care.

The goal of an EMS system is to provide organized, rapid, quality care to persons suffering from sudden injury or illness. The fire service is most often concerned with providing either all or a portion of the care that the patient receives between the time of the incident and arrival at a medical facility. Other elements of the EMS system such as private ambulance services, hospital emergency room personnel, municipal safety officers, and law enforcement personnel provide additional services necessary to maintain EMS as a system (Figure 7.2).

The idea behind the system approach is that by providing care through a coordinated system all potential resources can be utilized in the most efficient manner possible. This allows for better information exchange, more uniform techniques over a larger area, and a more effective method of analyzing present practices and procedures to determine if and where improvements can be made.

DETERMINING THE NEED FOR EMS

The level of emergency medical services a fire department needs depends on a number of variables, the most important of which are the needs of the community or area being served. To determine community need, the department must evaluate the answers to a number of questions and study the experiences of other departments with EMS programs. Then the department can de-

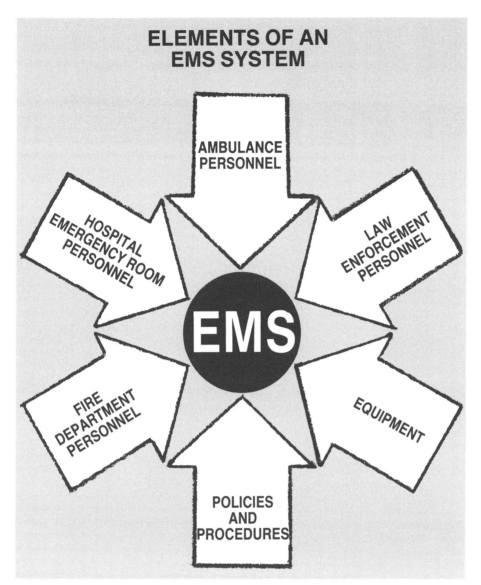

Figure 7.2 An EMS system is composed of many parts: all must work together smoothly to provide quality care to victims of sudden illness or injury.

cide what services it should provide. Some questions to consider are as follows:

- How many emergency responses are made in a given period? The number will vary with several factors, including population density, the size of the area being served, and the public's knowledge of the services provided. If an area has a population of 500,000, there will probably be more EMS responses than in an area with a population of less than 10,000. If the public does not understand the kind of emergencies EMS handles, some calls may come from people who want medication delivered from a pharmacy or who want to be taken to a doctor's office for a routine appointment.

- What types of EMS runs occur frequently? Departments can use this factor to determine what level of care, what specialized services, and what training they should provide. A substantial number of runs for cardiac emergencies indicates a need for advanced cardiac care units. If a community has a large number of auto accidents, the department should consider training in advanced trauma care and extrication (Figure 7.3). When an area has a very low number of trauma or cardiac incidents, the department may not need a high level of EMS. In any case, the department should perform a needs assessment before making a decision.

Figure 7.3 When automobile victims are trapped, special training is needed to remove them without causing further injury.

- How quickly do victims get emergency care after a call for help? Departments have to consider the response time to the scene of an incident and the travel time from the scene to a hospital. In remote areas where access is difficult or time consuming, EMS units may have to provide a higher level of care to keep patients alive until they get to a hospital. In a sparsely populated area where it takes a long time to reach a hospital, the department may also have to provide a fairly high EMS level.

- What services does the public feel it needs? Public opinion is an important factor. If citizens feel that present services are inadequate, they may express themselves through editorials, demands for legislation, or by their attitude toward EMS. Once citizens call for more advanced EMS, the department can hardly afford to disregard them. By the same token, it is hard to continue services at a high level if citizens feel the services are impractical or too expensive. A forced cutback is rare, however, because people can usually see the value of a good EMS system.

- What is the political opinion of need? Political opinion often matches public opinion. If a community wants a certain level of EMS or specific services, local lawmakers may follow their constituents' suggestions with legislation, ordinances, or codes. Political opinion also influences decisions on what organizations will provide the services, whether the jurisdiction will initiate pilot or experimental programs, and how the jurisdiction will fund training, operations, or equipment purchases.

- What services are being provided by what organizations? How many hospitals are in the area, and how are they distributed geographically? Do they provide specialized facilities such as trauma centers, cardiac care units, or burn centers? Does the nearest hospital have 24-hour emergency room staffing or is the emergency room open only during normal business hours? If a hospital staffs its ambulances with intensive care nurses or physicians, the department may decide not to duplicate the effort. The department must also consider the services local ambulance companies provide. Personnel with lower EMS ratings, whether on the fire department or with a local ambulance company, must be cross-trained to assist personnel with higher ratings.

- Who should provide the appropriate level of EMS to satisfy community needs? EMS is a system composed of several organizations that provide emergency medical services to the public. Each department must decide how it will cooperate with other community organizations, and what services it will provide.

The questions posed here are not the only ones that can be used to make a decision. Additional factors will vary from city to city or state to state. All must be thoroughly reviewed by fire department officials, hospital administrators, ambulance service operators, EMS commissioners, public health officials, and other interested persons. Once the decision is made, the community can begin to provide the funds, train the personnel, and implement the program.

LEVELS OF EMS

The EMS levels a department provides will depend primarily on how much training personnel have received, but there are other factors, too. For example, does the department transport patients? How much and what kind of specialized equipment does the department carry on its apparatus? Training also varies. What type of hospital facilities are available? A state emergency medical services commission or a local department of public health usually regulates the training program, certification, and policy for a given area. Even different programs provide three typical levels of training: first responder, emergency medical technician, and paramedic.

First Responder

First responder training courses are currently set up under the U.S. Department of Transportation. Formerly called "Crash Injury Management for Traffic Law Enforcement Officers," the course is more in-depth than a basic first aid course. It is designed to teach a person who would be first on the scene of an accident or injury how to handle any life-threatening situations with limited medical supplies. The course is oriented toward assessing and giving initial treatment to auto accident victims, but includes instruction on emergency childbirth, burn care, cardiopulmonary resuscitation, and poison treatment (Figure 7.4).

First responder courses involve fewer hours of instruction than an emergency medical technician course, the next level of training. DOT suggests that first responder courses require at least 40 hours of classroom and hands-on training.

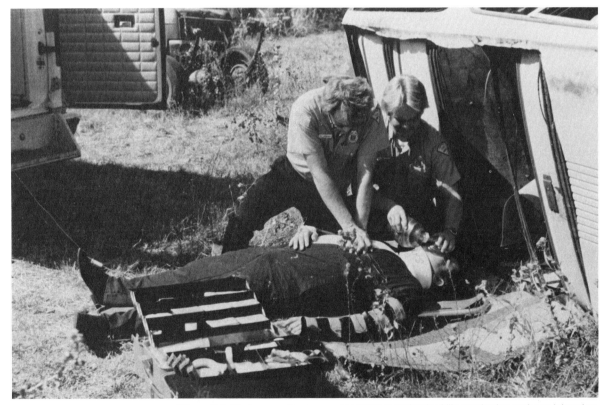

Figure 7.4 First responder courses teach the most basic skills required to assist an injured or sick person. *Courtesy of Glenn Pribbenow.*

EMT-Basic

EMT-Basics, trained in the classroom and hospital, are qualified in patient assessment, basic life support, and treatment of shock and other life-threatening conditions. Students also learn how to handle wounds, fractures, strokes, and childbirth. They also learn techniques of simple extrication, defensive driving, report writing, and incident scene control.

The training program used most often is one developed for DOT's National Highway Traffic Safety Council. The course requires at least 71 hours of classroom instruction and 10 hours of in-hospital observation and instruction. Hospitals, vocational training schools, or fire departments usually conduct the course using physicians, nurses, EMT-Paramedics, and other medical professionals and instructors.

Fire department operating ambulance or rescue squads often require personnel to complete at least an EMT-Basic course before assignment or during probation.

EMT-Intermediate

EMT-Intermediates complete the basic course and receive training in more advanced skills such as the use of anti-shock trousers, intravenous therapy, and advanced airway maintenance. The intermediate level is a fairly recent development for communities that need special skills beyond those taught in a basic course, but not as advanced as those taught in a paramedic course. The areas in which training is received and the amount of training depend on the need of the area being served. Some of the more common EMT-Intermediate courses are EMT-IV, EMT-Defib, and EMT-Cardiac (Figure 7.5 on next page).

Paramedic

Perhaps the most publicized level of emergency medical services is the paramedic level. Paramedics complete a thorough, intensive training program covering nearly every phase of emergency medical procedures including trauma patient assessment, cardiac life support, intravenous therapy and medication, and advanced airway maintenance.

DOT has developed a 15-module paramedic training curriculum that has become the most widely accepted program in the nation. A few states have developed programs for their own special needs through medical schools or county health offices.

This type of training is the most expensive and requires the most coordination between the fire service and the health care community. Paramedics must be trained by physicians and other health specialists, pass a state exam before certification, and are supervised by physicians responsible for the emergency health care program. These same conditions also apply for the more advanced EMT levels.

In addition to acquiring more advanced lifesaving skills, paramedics and advanced EMTs are able to use telemetry equip-

LEVELS OF TRAINING FOR PREHOSPITAL CARE

FIRST RESPONDER

Advantages:
1. It is a good, inexpensive and quick way to begin providing some EMS service if you have none. Also, many EMT programs are designed to build on first responder training. To go from first responder to EMT takes only an additional 40 hours rather than a total of 80 hours on top of the first responder's 40 hours.

2. It may be all the training you need in an area where another service (municipal or private) provides most of the EMS (as in parts of Kansas, for example).

3. First responder may be appropriate basic training in a fire service that provides tiered response, basic EMT ambulance and advanced life support paramedic units.

4. This level does not require physician involvement during training or certification.

5. Certification is recognized nationally.

Disadvantages:
1. It is the most basic training, just the minimum to get by with in truly life-threatening situations.

2. In non-life-threatening, but serious situations (such as a bad fracture) the first responder is left with no skills that are of help to the patient.

EMT

Advantages
1. Still fairly quick training with the added benefits of better patient assessment and more skills in managing a greater variety of injuries and other medical crises.

2. Absolutely required prerequisite training for intermediate and paramedic training.

3. EMTs provide better support than first responders in advanced life support situations that are being handled by paramedics or intermediates.

4. Minimum training necessary to provide a transporting EMS service, even one with only basic life support.

5. Taught by a non-physician and does not require physician approval for certification.

6. Nationally recognized certification.

7. Nationally accepted textbooks.

Disadvantages:
1. Since there are more skills to learn, there are more skills to lose if they are seldom used or practiced.

2. Cannot administer life-saving medications or fluids.

3. In an era of heightened public awareness due to television programs, and without adequate local public education, citizens may feel they're not getting what they've paid for.

4. Does require more training than first responder. May be more than necessary in a system where most EMS is delivered by another provider.

EMT-IV

Advantages:
1. Has all of the advantages of EMT plus the ability to provide fluids to patients in shock or where shock may be imminent, especially with trauma victims.

2. Particularly useful in rural areas where travel times to medical facilities are long, but resources, financial or otherwise, for training are low.

Disadvantages:
1. Can't give medications or defibrillate.

2. Skill degradation may occur if the personnel have too few opportunities to use or practice their skills.

3. Usually very limited background information is taught and personnel do not have a complete understanding of the physiological basis for their skills, which could have an impact on their judgment.

4. Requires development of specific protocols for the administration of IVs.

5. Requires physician involvement in training and certification.

EMT-EKG/DEFIB

Advantages:
1. Has all the advantages of EMT plus the ability to recognize certain life-threatening cardiac arrythmias on the EKG and to defibrillate if necessary.

2. Good training in areas with a high incidence of cardiac crises for quick responders where advanced life support EMT-Cardiacs or paramedics may be delayed.

(This level of training/certification is beginning to have a positive effect on survivability following cardiac arrest in some areas of Washington state.)

Disadvantages:
1. Can't start an IV or give medication to begin to correct the problem that caused cardiac arrest in the first place.

2. Requires physician involvement in training and certification.

EMT-CARDIAC

Advantages:
1. EMT plus IV plus defib plus the ability to give a variety of cardiac medications.

2. Cardiac-EMTs have a much better understanding of the disease processes with which they are dealing, especially those involving the cardiovascular and respiratory systems.

3. Can provide a basically paramedic service without as much of an investment.

4. Especially useful in urban areas with a high middle age population where more people are likely to succumb to cardiac problems.

Disadvantages:
1. Often, people trained and certified at the EMT-Cardiac level begin to gradually lose their basic EMT skills.

2. Skill degradation is a problem if EMT-Cardiac skills are not used or practiced routinely.

3. Compared to EMT and the other two intermediate levels, EMT-Cardiac is a considerable investment, even at the lower end of the training hours scale.

4. EMT-Cardiac requires physician involvement in training, certification and the establishment of operational protocols.

5. Since almost no EMT-Cardiac training standardization exists between states, the ability to hire pretrained people is restricted (and mobility is decreased for people within the system).

6. No nationally accepted textbook(s).

7. Personnel should have malpractice insurance coverage.

PARAMEDIC

Advantages:
1. Top-of-the-line training in terms of the maximum knowledge and skills accepted for prehospital providers.

2. If taught, paramedics have those rarely needed, but sometimes critical skills.

3. EMT-paramedic is a nationally recognized level of advanced life support training and certification.

4. A national organization exists to provide the certification examination which saves the state the work of developing one.

Disadvantages:
1. Skill degradation can be a serious problem.

2. Tremendous investment of resources.

3. Requires physician involvement in all phases; training, certification and operational protocols.

4. No nationally recognized textbook (though that situation is improving).

5. Even with the national standard, considerable variation still exists across states in terms of the optional skills taught.

6. In states that do not provide their own certification examination, testing is done by a private organization, which sometimes can become a political issue.

7. Personnel should have malpractice insurance coverage.

From "Do We Really Need Paramedics?" by Dorothea R. St. John, printed August 1984 issue of Fire Command. Reproduced with permission.

Figure 7.5 The three levels of training for emergency response are first responder, EMT, and paramedic. Departments must assess the needs of the community as well as their own resources to decide what levels of care to provide.

ment to transmit a patient's electrocardiogram from a remote location to a physician at a hospital. The physician uses the data to diagnose the patient's condition (Figure 7.6).

Telemetry equipment picks up minute electrical impulses from the body, modulates them into an audible signal, then amplifies and transmits the signals as radio waves. A console in the hospital emergency room receives the transmission and converts it into an audible signal. The console also converts the signal into a visible output on a screen and a recording device so physicians can keep a record of the patient's heart activity. Most models provide voice contact between the paramedics in the field and the physician so they can exchange additional information.

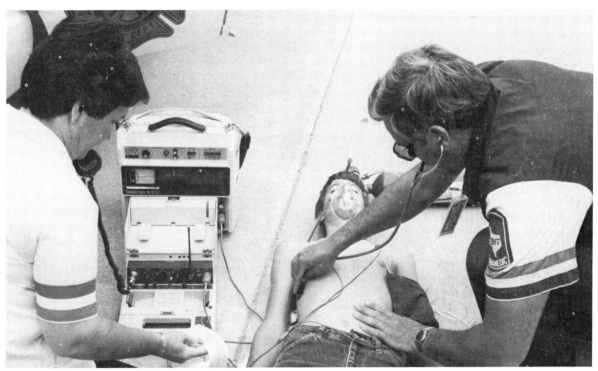

Figure 7.6 Telemetry equipment uses radio waves to transmit a patient's heart signals to a physician in a hospital. *Courtesy of Colbert, Oklahoma Fire Department.*

EMS CERTIFICATION

After receiving initial training, personnel are tested and then certified by state EMS commissions, the National Registry of Emergency Medical Technicians, or the authority having jurisdiction in the area. When personnel are certified by an authoritative agency, there is a greater assurance that quality care will be received by the patients. Some states require EMS personnel to be certified or licensed before they can provide emergency care on a regular basis. Any employees whose job requires them to give emergency care fall under these licensing or certification provisions that require EMTs to conform to standards of care equal to those provided by others at the same level of training. This helps maintain a high quality of emergency care, which is the best way to avoid lawsuits, and is the only acceptable way to operate any level of EMS. Fire departments should actively lobby for such a provision if their state does not have one.

Certified emergency medical providers are required to obtain recertification, generally at two-or three-year intervals. During the validity period of the certification, continuing classes are attended in such areas as cardiopulmonary resuscitation, patient assessment, airway maintenance, and extrication (Figure 7.7). The total number of hours of training and the allowable subjects for continuing education vary with the certifying agency and the level of training that has been received. For example, a paramedic will generally require more hours continuing education than will an EMT-Basic.

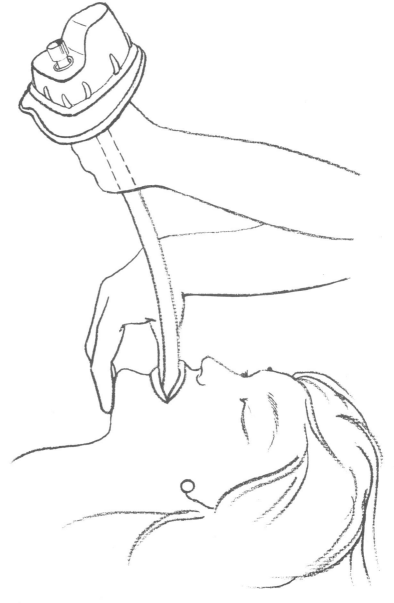

Figure 7.7 Emergency medical providers must attend continuing education classes in such skills as cardiopulmonary resuscitation and airway maintenance.

In addition to the continuing education classes, a refresher course is required. The refresher training is received during the final year that a certification is valid and is a condensed version of the original training program. Once these requirements are completed and the individual successfully completes an examination, recertification may be granted.

The process of recertification helps keep the EMS personnel up-to-date about medical advances and changes in practices or procedures. It also ensures periodic practice of skills that may not be used as frequently as necessary to sustain proficiency.

The application certifying agency should be contacted for specific requirements regarding recertification.

SUPPLEMENTAL EMS ASSISTANCE

Several communities supplement their emergency medical services with agreements with other jurisdictions and agencies. Some communities have developed Military Assistance to Safety and Traffic programs that use military helicopters and active duty medical corpsmen. Communities benefit from the additional air ambulances and emergency personnnel available, and the military units receive realistic instead of simulated training. Although civil authorities and the base commander usually work out the scope of the local program, there are some restrictions. No military personnel or equipment can be relocated to activate a MAST program, and existing programs cannot interfere with the primary military mission. Only military training funds can be used to support the program. Other assistance might be attained through mutual aid agreements with other fire departments, governmental units, or outside resources. This can be critically needed in the event of a major disaster.

GOOD SAMARITAN LAWS

For many years, people who gave first aid at accident scenes were vulnerable to lawsuits that alleged improper care. As a result, many states enacted "Good Samaritan" legislation to protect them. This relieves them of legal responsibility for any errors or omissions in the care they provide. The legal provisions differ from state to state and do not grant absolute immunity in cases of gross negligence or willful and wanton misconduct that injures a patient. Often the Good Samaritan laws apply only to services rendered at the scene and not to those rendered while the patient is being transported to the hospital. Some laws apply only to dentists and physicians who chance on an accident scene, some protect emergency medical technicians, and others protect anyone in the state who gives help.

Individual EMTs may be protected from lawsuits, but not necessarily the fire department. Volunteer departments may incur different liabilities than career departments. It is extremely important to be well informed about the laws in your state; also know whether and to what extent your current insurance policy covers EMS- related cases.

MANAGING EMS

There are a number of roles EMS personnel may play in a fire department depending on the location and work load of the department. They may function as dual role, cross-trained firefighters, career EMT-Paramedics, or civilians who are trained only to provide emergency care.

Firefighter/EMTs (Cross-Trained)

These people have completed full training in suppression and are also trained either as EMTs or paramedics (Figure 7.8). Often they must first complete a probationary period as firefighters before they are trained in emergency medicine, although some departments hire EMTs first and then cross-train them in suppression. Others cross-train the individuals immediately before they graduate from recruit academy.

Figure 7.8 Many fire departments cross-train firefighters in suppression and emergency medical skills.

Cross-trained personnel are seen by many as the most logical and cost-effective way to integrate emergency medical services into the fire service. The proponents of this method cite the following advantages:

- Start-up costs can be lower because a fire department already has in place a building to house emergency vehicles, a dispatch/communication system, and personnel trained to respond to emergencies.
- The selection process is simplified because the firefighter is familiar as an employee.
- The fire department is seen as more cost effective because there is less down time for fire fighting personnel and citizens are getting more for their tax dollars.
- There is greater flexibility in assigning job duties. Personnel can be rotated through suppression and EMS assignments because all possess the necessary skills.
- The firefighter who suffers from burnout can transfer back to suppression.
- Firefighters have the opportunity to advance up the department career ladder.
- Because of the larger number of emergency calls, the department may be able to provide pay incentives for EMS personnel.

There are, of course, disadvantages to this system. Departments that handle a very high number of fire and medical calls may be unable to use firefighters in both roles. Because of the higher incidence of EMS calls, personnel may burn out from overwork if work loads are not distributed carefully. Pay differentials may cause resentment among non-EMT firefighters. Also, firefighters must perform both EMS and fire fighting duties regularly to keep skills current. Some departments rotate personnel through EMS and various suppression duties to alleviate this problem. Others conduct regular in-service training sessions.

Some departments use their EMS personnel only in limited suppression roles while others use them for full suppression roles and rely on other units to handle any EMS calls that may occur.

Single-Function EMS Personnel

These people are hired by the fire department to handle emergency medical calls only; they are not trained as firefighters. There are several reasons for employing this method of providing EMS:

- The fire department does not want to involve suppression personnel in emergency medical care. This arrangement may be attractive if the department is able to hire extra people for providing EMS.

- The department responds to such high numbers of both fire and medical emergencies that it is not possible to blend roles.
- The feeling that suppression and EMS are different specialties and cannot be performed adequately by one person.

Maintaining separate fire fighting and EMS personnel requires strong leadership to prevent tensions. Since firefighters and EMS personnel usually respond from the same building, the differences in the two groups' work or work loads will be highlighted. This can lead to friction: Firefighters may feel that EMS crews disrupt station life because of the higher frequency of medical calls. The EMS crews may feel that they are looked down on by firefighters.

The attitude communicated by the chief is critical: The message, spoken or implied, that EMS is a bother is very detrimental to the morale of EMS personnel (Figure 7.9). Employees who feel that they are "stepchildren" and have difficulty getting the chief's attention will have a hard time maintaining pride in their skills and jobs.

Figure 7.9 Departments that maintain separate fire fighting and EMS personnel must be carefully managed to avoid tensions.

Department personnel who are familiar with the duties of "the other side" and know the importance of each division to the fire department will likely have greater respect for each other's skills. Responding an engine to an EMS call or sending EMTs to a fire scene for the safety of the firefighters will give everyone some street education. If possible, cross training should be considered as a means of providing more flexibility when assigning duties.

The lack of a well-defined career ladder is another problem area for single function EMS personnel. Larger departments need rescue team supervisors, senior paramedics, and EMS coordinators. In smaller departments there may be no promotional opportunities at all, leaving the EMS employees nowhere to go but out of the department. Many people assert that more administrative positions that make use of an employee's medical and supervisory skills need to be created.

Transportation of Patients

Is it to the department's advantage to transport patients or to allow a private or municipal ambulance service to take over? Some departments prefer to transport because it allows them to have complete control of the patient's prehospital care. Others prefer to leave transporting to an ambulance service, even though this requires that ambulance personnel be briefed on the patient's condition, and may even necessitate one EMT or paramedic riding in the ambulance. In that way, the time the EMS company is out of service is reduced. All EMS providers must handle system abuse: people calling who do not really have medical emergencies or who are looking for a free ride to the doctor. Whoever provides transportation is often placed in the awkward position of trying to convince people to call for private transportation instead of the "free" fire department service.

STRESS AND BURNOUT

Departments that provide EMS respond to two or three times as many emergency medical calls as they do fire calls. That and other reasons make the job of the EMT-Paramedic more stressful than that of a firefighter. Some of the leading causes of stress for EMS personnel are

- Overwork
- Lack of sufficient training
- Friction with firefighters
- Poor management — difficulty dealing with chief, hospital personnel
- Dealing with injury, death, and distress of survivors
- Abuse of system — having to respond to many calls that are not really emergencies

- Lack of promotional opportunities
- Continuing education and recertification

If these stressors accumulate, personnel may suffer a syndrome known as burnout. Personnel suffering from burnout show an emotional withdrawal from patients and/or coworkers, lowered interest in skills or continuing education, mood changes, increased illness or absenteeism, and increased use of alcohol, food, or drugs.

Of course, it is to a department's advantage to try to lengthen the career of any well-trained firefighter or EMT-Paramedic. There are a number of steps that can be taken to alleviate stress and prevent burnout:

- Organize regular sessions to discuss ways to improve EMS, cope with stress or air gripes. Often people just need someone to talk to.
- Provide access to counseling if necessary. This must be kept strictly confidential in order to demonstrate a sincere commitment to help and to avoid placing additional stress on the individual.
- Limit the length of duty. Some departments place EMS personnel on a 12-hour shift rather than a 24-hour shift.
- If the individual is also a firefighter, consider a transfer back to an engine or ladder company.
- Increase time off between shifts.
- Rotate individuals to stations that handle a lower number of calls.
- Maintain good communications between hospital and EMS personnel to reduce confusion about duties and operating procedures.
- Encourage regular exercise. Look to the local exercise facilities or university health programs for assistance.
- Help educate the public to reduce nonemergency calls.

Fire Communications Systems

8

CHAPTER 8 OBJECTIVES

1. Identify communication breakdowns and suggest or implement procedures or equipment changes.

2. Evaluate hiring and training procedures for dispatchers; implement changes to increase efficiency of service.

3. Identify individuals or groups in the community who have special needs with respect to emergency services; develop procedures for providing efficient emergency response.

4. Analyze the cost effectiveness of upgrading all or part of the communications system, including start-up costs, maintenance and repair costs, and ability of system to be expanded.

5. Evaluate the communications system for efficiency of integration with other emergency services.

6. Determine emergency backup systems needed to continue emergency services in the event the main communications system malfunctions.

Chapter 8
Fire Communications Systems

A fire department's communications system is an essential part of any community's fire defenses. A good system speeds response and plays a key role in guiding department activities. The system has three main functions. First, it receives alarms from the public and directs them to the appropriate fire fighting units and personnel. Second, it is a method of exchanging information between personnel at the scene and the communications center. Third, superiors use the system to get information from subordinates and then direct fireground operations. The system is divided into alarm and radio components.

ALARM SYSTEMS

A fire alarm system is composed of municipal fire alarms, private alarm systems, and private and public telephones. The fire alarm system receives an alarm from the public and retransmits it to the appropriate fire companies and personnel.

Municipal fire alarm systems are classified by how they operate: telegraph, radio, telephone, or a combination of these. NFPA Standard 1221, *Public Fire Service Communication,* classifies these systems as Type A or Type B. The Type A system receives an alarm signal from a box and manually resends it from the receiving station or communications center to the fire stations. The Type B system automatically retransmits the alarm signal.

Public Telephone System

The public telephone system is the most widely used method of transmitting fire alarms. In many areas, such as outlying suburbs or rural settings, it is the only method of rapid communication. A major advantage is that the dispatcher or operator can ask the caller about the nature of the emergency, and obtain the address or the call-back number. Telephone systems with Enhanced 911 display the caller's number and address automatically. Some 911 centers incorporate telecommunications devices for the deaf (TDD) which enable deaf persons to contact the 911 center using a keyboard.

Telephone lines are connected to a recording device in the alarm center. If the caller hangs up or is disconnected, the information received can be played back. The recording device also is important when callers are so excited that they cannot be understood, or when they speak a foreign language (Figure 8.1).

Figure 8.1 Stickers can be placed on the telephone to assist callers when under stress.

Telephone accessibility can be a problem in some areas. In business districts, offices and stores might be closed at night, on weekends, or holidays. In resort communities, many buildings might be closed and locked during the off-season. There are more outdoor telephone booths than ever before, but coin-operated units are useless if the caller does not have the correct change. To remedy this, callers in many cities can reach the operator without using a coin.

Telephone circuits will not work under some conditions. Storms or other natural disasters can knock down phone lines, or the fire itself can destroy the circuits. People might not know the phones are out of order until they try to call for help in an emergency. Some systems can handle several calls at once, but the lines are swamped during an extreme emergency. Other systems can receive only one call at a time, and sometimes two or more people try to call at the same time. The telephone system cannot preempt a routine call when an emergency number is dialed. If the line is busy, the caller must wait until the line is free or try to send the alarm another way.

A telephone service area will often include more than one fire department with separate emergency telephone numbers. This can and does cause delayed or lost alarms, or incorrect fire company assignment. People who change addresses frequently might not take the time to find out the fire department's emergency number or even what fire department serves their area unless they have a fire, consequently causing delayed reporting. When one telephone central office serves more than one community, an alarm might go to the wrong department. Firefighters will arrive and find no fire while in a neighboring town a structure burns at the same street address.

The nationwide 911 emergency telephone number has been adopted in large and small communities. The three-digit number has many advantages over the standard seven-digit telephone number. It is easier to remember one number for any emergency than to remember one for police, one for fire, and still another for ambulance service. The number can be dialed in the dark with little problem. If the telephone service area incorporates more than one department, alarms can still go to the wrong fire department even with 911. Some systems solve the problem with a selective router that identifies the caller's location.

Telegraph Systems

The first municipal fire alarm system, developed in 1847, was a telegraph-type system. Modern telegraph systems have changed little since those days. Depressing a lever or handle on the box releases a spring-wound clockwork mechanism that rotates a code or programming cam that opens and closes a set of electrical contacts (Figure 8.2). The contacts work like a switch to energize and de-energize the alarm circuit and transmit a coded message. The code is the box number of the sending unit, so dispatchers can use the number to determine the location of the alarm.

In the communications center, the box circuit is connected to the receiving equipment. When the alarm is received, it is printed or punched on a paper tape and the bell sounds. At the end of the transmission, the time is stamped on the tape by hand or automatically.

Older telegraph systems had a major drawback: If more than one box on the same electrical circuit was pulled at the same time, their messages came in at the same time. If the communications center received them at all, the signals could be garbled or the box number wrong. Modern systems have "noninterfering" and "succession" features. If up to four boxes are pulled at or about the same time, the first one will transmit and the others will run but

Figure 8.2 The telegraph fire alarm box has a coded wheel that transmits a signal when activated.

not transmit. Once the first box has transmitted, the other boxes will transmit in succession. These two features greatly reduce the potential for receiving garbled messages or none at all. A series box circuit can also carry signals from telegraph and telephone-telegraph units. The latter unit is just a telegraph box with a telephone handset.

Telegraph systems only transmit a number; this feature is considered by many to be a disadvantage. Firefighters must at least get close to the scene before they know what kind of alarm they are answering. Telegraph systems are vulnerable to false alarms.

Telephone Fire Alarm Systems

This type of unit is essentially a telephone handset mounted in a specially designed housing (Figure 8.3). Each handset is connected to a communications center switchboard. When the handset is lifted off its cradle, a lamp on the switchboard lights up to indicate that an alarm is coming in and to identify its location. Some telephone systems are equipped with a concentrator-identifier. This concentrates a number of individual box circuits at one location and transmits a signal from there to the fire alarm

Figure 8.3 Telephone-type fire alarm boxes are used for direct voice transmission to the communications center. Lifting the handset indicates the box location.

switchboard over a smaller number of circuits. The box used is identified on the switchboards with a direct circuit telephone.

Telephone systems are arranged in parallel or series circuits. For each alarm box, the parallel circuit has a pair of wires connected to the switchboard. Large departments will use concentrator-identifiers to reduce the number of terminations. About 200 box circuits can be terminated at the concentrator, which is in a telephone building. From the concentrator, ten tie lines extend to the communications center switchboard. In effect, the equipment concentrates the number of street boxes served by one termination with a minimum of two tie lines. A box circuit connected in series serves several street boxes with the two wires, reducing the number of switchboard terminations. Lines from the concentrator equipment do not serve specific boxes but are responsible for the first ten alarms received. A holding arrangement stores additional alarms until a tie line is free. The identifier in the communications center trips visual and recording equipment to identify the box transmitting the alarm.

Radio Systems

Radio systems also transmit coded messages using radio waves instead of wires and electric current (Figure 8.4). The transmitters are powered by batteries, prewound generators, or

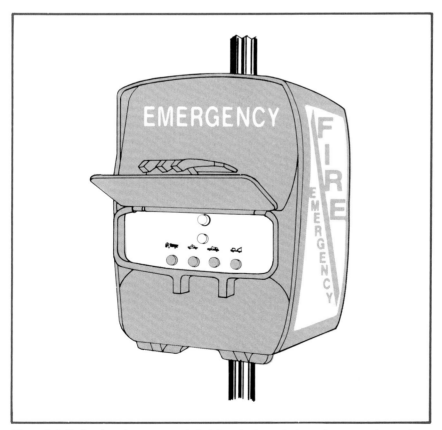

Figure 8.4 Radio alarm boxes transmit their location by an independent transmitter and are not restricted by wired circuits.

user-powered generators. Battery-powered units transmit periodic signals to indicate that they are working or send another signal to indicate that the batteries are low. The batteries on some units must be replaced periodically, but others have solar recharging capabilities. Prewound generator units get power when the user releases a latch so a mainspring can rotate the generator. In the user-powered units, the user pulls on the handle to rotate the generator.

Radio systems have no need for noninterfering or succession features because the alarm signal is transmitted so quickly that two signals seldom overlap. Most boxes are also designed with random spacing between the required three sets of alarm signals. When multiple boxes are tripped at the same time, the communications center receives at least one correct set of signals from each box.

RADIO COMMUNICATIONS SYSTEMS COMPONENTS

There are three basic systems fire departments use for radio communications:

- Single-frequency simplex
- Two-frequency simplex
- Two-frequency duplex

A simplex system can transmit broadcasts between two users only one way at a time; that is, only one user can be on the air at a time. The single-frequency simplex system is the two-way radio system used most widely. It is the most inexpensive to equip and takes up the least space in the radio spectrum. One advantage of the system for the fire service is that all units responding to an alarm can hear all the broadcasts dealing with the emergency. In the two-frequency simplex system, central communications broadcasts to mobile units on one frequency, and the mobile units respond on another. The system has at least one advantage for nearby agencies using the same frequencies: the broadcasts of one base station do not interfere with the mobile unit broadcasts of another agency. A disadvantage is that mobile units cannot receive transmissions from other mobile units. A duplex system allows central communications to broadcast and receive transmissions at the same time.

Departments planning a communications system should strongly consider one with several frequencies. These systems are more flexible than their single-frequency counterparts. For instance, if several frequencies are available, the department can assign the incident to a frequency not used for dispatch. Departments usually reserve one frequency for outgoing messages and another for incoming messages. Mobile units then receive only the messages they have to act on.

One other element of a communications system is the capability to alert or contact individual department members, usually with some form of pocket pager. Pagers are small, battery-operated radio receivers about the size of a pack of cigarettes. One type is silent until it receives the proper alerting tone from the base station; then it activates to give a message. The other type gives only an alarm signal when activated. Pagers can also be equipped with a monitoring signal that allows wearers to hear radio traffic about the emergency to which they are responding or working.

RADIO FREQUENCIES

In the early days of radio, transmitter owners broadcast on any frequency they chose. As the number of broadcasters increased, the situation became more and more chaotic. Stations would interfere with each other with the signals and broadcasts "bleeding" into one another. Stations with more power could override the signals of weaker stations. Eventually, the forerunner of the FCC was created to regulate use of the radio spectrum, which was split into frequencies based on radio wavelengths. The system has evolved to the extent that today the FCC assigns specific frequencies to successful broadcasting license applicants.

The FCC has set aside four bands, or groups of frequencies, for land mobile service, of which fire safety operations are a part: 50 MHz (low band), 150 MHz (VHF), 450 MHz (UHF), and 800 MHz. In any given location, the fire service can be assigned to any combination of these frequencies, depending on the department's needs and how crowded the airwaves are. Historically, fire departments were assigned low-band frequencies until the low band was filled with users. Then the FCC began assigning the fire service increasingly higher frequencies in that specific area.

The FCC also has reserved channels for mutual aid, administrative use, and emergency medical services. This frequency allocation changes yearly. Generally, the FCC reserves low band for rural use because low-band signals travel farthest. VHF has a stronger signal with a shorter range than low band so the FCC generally assigns it to departments in urban areas or in combined urban-rural areas. UHF goes to departments in the larger metropolitan areas because this band combines a stronger signal with a shorter range and works well where tall buildings might interfere with broadcasting.

The 800 MHz spectrum and enhanced trunking offer a wide range of options for agencies needing additional frequencies for emergency operations. It provides excellent transmit/receive coverage and superior penetration in such problem areas as tunnels, subways, and congested city areas with tall buildings. Enhanced trunking makes the most efficient use of shared frequencies by automatically assigning channels by priority: life-threatening emergency broadcasts, tactical situations, command, operational, and normal broadcasting. Radios do not even have to be set to the correct frequency since the system automatically selects a clear frequency for both transmitting and receiving. Channels can be used for data transmission and telephone interconnect, allowing vital technical information to be transmitted to chief officers. The report of a government/private industry team, "900 MHz, Trunking Communications Systems Function Requirements Development," is available from APCO, 105½ Canal Street, New Smyrna Beach, FL 32070 for $10.

Fire departments have indirect use of another set of low-band frequencies — the citizen's band. Local CB clubs are often organized to provide emergency assistance, especially those affiliated with REACT (Radio Emergency Associated Citizens Teams), headquartered in Chicago, and ALERT (Affiliated League of Emergency Radio Teams), headquartered in Washington, D. C. Some clubs will listen to CB channels and contact public safety organizations about distress calls they have monitored. Other clubs have members trained in traffic control, search, and first aid. However, departments should not let club members handle dangerous fire service duties for which they are untrained.

FEDERAL REGULATIONS

The highest authoritative body governing broadcasting in the United States is the Federal Communications Commission. The FCC grants licenses for radio use, assigns frequencies for transmitting, designates call letters, determines the maximum power a transmission station may have, and determines the rules and regulations for operating radio equipment. Similar agencies and organizations regulate broadcasting in other countries.

The FCC requires operators to keep a radio log including the time and nature of all transmissions. A typical series of entries might read as follows:

- 1827 hours: Alarm box 263, Engine 12, Engine 9, Ladder 6, Battalion 2 assigned to 3723 E. Main, Sue's Flower Shop
- 1829 hours: Engine 12 on-scene, light smoke visible
- 1830 hours: Dispatch call letters
- 1831 hours: Battalion 2, Engine 9, Ladder 6 on-scene
- 1844 hours: Battalion 2 transmitted control of fire, placed Engine 9 in-service
- 1857 hours: Battalion 2, Engine 12, Ladder 6 in-service, returning to quarters
- 1901 hours: Battalion 2, Engine 12, Ladder 6 in-quarters
- 1902 hours: Engine 9 in-quarters

Fire departments should use radio recorders even though they are not required by FCC regulation. Recorders document all radio traffic and dispatching information to provide an accurate account of operations. They protect the department and its members when questions are raised about communications and operations; they also document such evidence as dispatch time and company arrival on the scene in case of litigation.

The recording devices run continuously or intermittently. The continuous type operates even when no transmissions are taking place; the intermittent units run only when traffic is on the air. Because they run all the time, continuous units use more tape and are more expensive to operate than intermittent types. Intermittent units can miss the beginning of a transmission because they are actuated when traffic is broadcast, and it takes a little time for the recording to begin. If an operator speaks before recording begins, the recorder misses the first part of the message. Operators can overcome the problem by using proper procedures, by pausing after keying the microphone and before speaking.

Private Alarm Systems

Private alarm systems are used in industrial plants, private residences, hospitals, shopping malls, warehouse complexes, and several other locations (Figure 8.5). Such systems, normally called protective signaling systems, can be used to do one or more of the following:

- Notify occupants to evacuate.
- Summon the fire department or other organized help.
- Supervise extinguishing systems to make sure they work.

Figure 8.5 Proprietary signaling systems are privately owned and maintained, and retransmit signals to the public alarm center.

- Supervise industrial processes to warn of abnormalities that could contribute to fire hazards.
- Supervise building personnel to make sure they perform their assigned duties.
- Activate fire control equipment.
- Identify the location of the fire.

The types of private signaling systems are

- Central station protective signaling systems
- Auxiliary protective signaling systems
- Local protective signaling systems
- Remote station protective signaling systems
- Proprietary protective signaling systems
- Household fire warning systems

These systems notify the fire department in different ways. In a central station or supervised system, manual pull stations or automatic detectors send alarms to an independent agency. When an alarm is received, trained agency personnel notify the fire department over direct lines.

Auxiliary system lines are part of the municipal fire alarm system. The alarm is activated and then sent from the protected premises over the municipal alarm lines to the fire department where an indicator lights up. The system receiver in the alarm room might also print out the alarm. Public buildings such as schools and hospitals are often protected by auxiliary alarm systems.

Local systems are designed to notify only the occupants of the protected location. Someone in the building must notify the fire department another way, such as by telephone. The remote station is similar, but it also relays the alarm to the fire department or a communications agency that in turn notifies the department.

Proprietary systems protect large plants or installations that have their own central supervising station from which the alarm goes directly to the fire department.

Household fire warning systems, like the local systems, generally only notify occupants. The fire department must often be notified another way, but sometimes these systems can be supervised.

Several new fire alarm systems have been designed and installed in recent years. Fire department managers should study all new systems before installation to ensure that they are compatible, effective, and manageable. Fire department management and companies that supervise alarm systems must set up procedures that clearly outline when and by whom the department will be notified when an alarm is received.

ALARM CENTERS

Alarm centers, also known as communications or dispatch centers, receive alarms and transmit signals to the fire stations to initiate the response of apparatus and personnel. A communications center can serve a single community, an entire county, or a group of adjoining municipalities; but if it uses any type of box alarm system, it should have these general capabilities:

- The center should receive and record alarms automatically, including the location and the time received.

- Equipment receiving the alarms should alert personnel visually and audibly, showing the alarm's exact location.

The communications center should also have equipment to receive supervisory and trouble signals. Trouble signals should actuate visual and audible alarms that personnel can easily distinguish from fire alarm signals. The audible indicator could have a reset switch, but the indicator must be able to reactivate if a second trouble signal from another box comes in.

Since municipal fire alarm systems must work at all times, departments should provide two power supplies for box and alarm circuits and related facilities. Power sources include public utility services, storage batteries, and emergency generators. The primary power supply, usually a public utility line, serves the center and its equipment under normal operating conditions. The secondary source provides power if the primary source fails. The secondary source, most often storage batteries or an engine-driven generator, should switch on automatically when the primary source fails.

In addition to the primary and secondary systems, fire departments can install a direct trunk line between dispatchers and fire stations to maintain communications when dispatching equipment goes out. The line is usually a separate unsupervised circuit. (See NFPA *Standard 1221, Standard for the Installation, Maintenance and Use of Public Fire Service Communications.*)

Departments also have to locate and construct the communications center so service will seldom be interrupted because of fires or other causes. The center should be built of fire-resistant materials and away from high-risk hazards. A single structure in a relatively open area is preferred but not always possible. If the alarm center is in the same building as the city hall or fire station, the center should be separated from the rest of the building and access limited. These measures prevent unauthorized entry, control the number of persons in the room, and protect the center from fire exposure, civil disturbances, and natural disasters.

Staffing

A system is only as good as the people who operate it. Fire alarm dispatchers should be healthy and free of any physical or mental defects that would affect their ability to efficiently handle their duties. They should be temperamentally suited for the job, able to remain calm and take decisive action during emergencies, able to remain alert during long periods of inactivity and repetitive operations, and able to work harmoniously with other people (Figure 8.6).

Figure 8.6 Fire alarm dispatchers must be able to remain calm during emergencies and work well with other people.

The dispatching job is a good one for civilians, but departments with civilian dispatchers might want the dispatch supervisor to be a sworn officer. Regardless of the dispatcher's status, an active training program at the entry level and beyond is critical to the system's effectiveness. A dispatcher's error in judgment can significantly affect the outcome of an emergency and might get the department sued.

Dispatchers should continue to increase their knowledge of the area in their jurisdiction, street names, and locations of high life-hazard structures such as hospitals, schools, and some industries. They should also have a working knowledge of the fire alarm system to test alarm reception and transmission equipment when necessary.

Dispatchers should know how to control voice tone and speed. They must also know local dispatch signals and codes and local, state, and federal radio regulations. Departments are moving away from the "10" codes to clear text in order to reduce misunderstandings during radio transmissions.

Because of budget considerations, many municipalities are consolidating services to use personnel and equipment more effectively. Some cities combine fire and police department dispatch and alarm centers. This increases dispatcher work load and frequently warrants increasing the number of dispatchers required. Dispatchers might handle from 10 to 20 emergency transmissions with each routine fire alarm. Add to this the typical telephone calls and routine police traffic, and the abilities of a dispatcher can be overloaded quickly even under normal conditions.

Combination fire-police alarm and dispatch centers provide a common point to receive calls on the 911 system. A person who dispatches for both police and fire departments must know police department dispatching procedures, policies, signals, and radio language for both agencies. The two departments can minimize the problem by standardizing their terminology.

Dispatching Procedures

One of the most critical periods for dispatchers is the time an alarm comes in, and they should be well trained to get the right information quickly to start units on their way. Skill is especially important when the public alerts the department by telephone. One former dispatcher gives the following tips on what dispatchers should be trained to do when a citizen calls in an alarm. They should

- Ask if there is an emergency and, if so, ask about the problem.

- Have their questions organized to control the conversation so they can get the information they need, and ask them in an assertive voice.

- Get the kind of information that lets them picture what type of emergency really exists.

- Ask the caller's name and the number called from.

- Make sure they have the exact location of the alarm, asking about cross streets and other identifying landmarks if necessary.

- Get a call-back number in case firefighters cannot find the alarm.

- Do not let the caller off the phone until they have all the information they need to dispatch responding units, or they are sure there is no emergency.

The answers should be recorded on some type of emergency alarm report (Figure 8.7 on next page). Then the dispatcher can get about the business of dispatching the responding units. Dispatchers must realize that dispatching delays can increase response time.

COMMUNICATIONS CENTER

FIRE DEPARTMENT DISPATCH REPORT

☐ CITY
☐ COUNTY
☐ BORDERLINE

__________ INCIDENT NUMBER

__________ MAP PAGE

DATE _____ CALL TIME _____ REPORTING PARTY _____ CALLBACK NUMBER _____

INCIDENT INFORMATION

TYPE OF INCIDENT	*DESCRIPTION*		
Fire			
Smoke Investigation			
Medical Aid			
Rescue			
Hazardous Condition			
Aircraft			
Alarm: Autom/Manual			
Public Service			
Needless Call			
False Alarm			
Mutual Aid			

__________ REPORTED LOCATION

__________ CROSS STREET

__________ ACTUAL LOCATION

ADDITIONAL INFORMATION __________

INCIDENT COMMANDER __________

UNIT #	DISPATCH TIME	AT SCENE	UNDER CONTROL	CANCEL	AVAILABLE	FINISHED ASSIGNMENT	IN QTRS

B.C.F.D. INFO	☐ County Call ☐ Mutual Aid Request _____ Notification Time	___ Engines ___ Hand Crews ___ Water Tenders	___ Chiefs ___ Dozers ___ _____

DISPATCHED BY __________ ASSISTED BY __________

Figure 8.7 All emergency calls should be recorded on an emergency alarm report. *Courtesy of Chico, California, Fire Department.*

Different departments have different ways of getting units to an alarm location. Some use a system of bells or other sounding devices, others use radio-voice communications, and still others have an automatic, computer-operated system. In Montgomery County, Md., for example, dispatchers record calls on the appropriate form, determine the alarm box area from a file, and use a running card index to dispatch the units that are supposed to arrive at the location first. After responding units are alerted by radio, a company member punches an acknowledgement switch to let the dispatcher know the alarm was received, without tying up the radio. Then the dispatcher updates an electronic status board that uses colored lights behind each unit's nameplate: red for call; green for on the air, out of the station, but available; yellow for out of service; white for transferred to another station, off the air. No light showing behind the unit's nameplate indicates it is in its assigned station ready for call.

Some fire departments try to get information about the location of the alarm, much of it from pre-fire plans, to fire companies as they respond. Some departments transmit the information by radio; others have the information available in individual fire vehicles on transparencies or microfiche. Some departments are "pre-alerting" their stations by transmitting the address of a call while researching the dispatch information. Units respond when the address is in their area. The system reduces response time.

Dispatching for Emergency Medical Services

Dispatching for EMS differs from fire response in that not only must the operator obtain the address of the emergency, but must also attempt to determine what type of emergency exists. Although many departments respond to all calls, the greater number of EMS calls (two to three times as many as fire calls) and often scarce number of ambulances has led many localities to institute a system of call screening and priority dispatching (Figure 8.8 on next page). Call screening is an attempt to weed out the large numbers of nonemergency calls in order to leave ambulances free to respond to real emergencies. It is also used to determine how high a priority should be given to the call. It is not an attempt to diagnose but rather to determine the severity of the injury or illness without being able to see the patient. Then, depending on the units available, the dispatcher must decide which units to dispatch or whether to refuse the call. Although many calls can be screened out, the possibility of a lawsuit exists if a call is refused and it turns out to have been a real emergency. Screening protocols must be followed strictly and the rule is: When in doubt, send an ambulance.

One common call screening technique is to ask several questions in order to categorize the call (head injury, for example).

Then the EMS operator asks an ordered set of questions that cover the most serious symptom to least serious symptom in that particular category of injury. Depending on the answers given, the dispatcher can immediately stop the questions and order an ambulance or engine company to the scene. Some departments use a flip chart for asking questions; others have questions programmed into a computer that can make recommendations as to what type of response should be made. The Department of Transportation's National Highway Traffic Safety Administration has developed a 25-hour course entitled Emergency Medical Dispatch Priority Training. This course teaches skills in caller interrogation, call screening, and pre-arrival patient intervention.

Survey of Eight Emergency Medical Service Systems

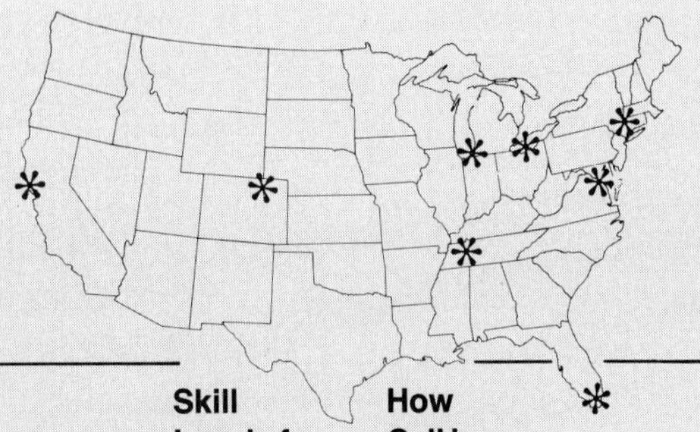

CITY	Part of Fire Dept. or Independ.	Type and No. of Units	No. of Calls/Year	Skill Level of Dispatcher	How Call is Handled
Nashville, TN	FD	12 ALS No FD response	50,000 calls/year	EMTs	Go on every call. Determine chief complaint.
San Francisco, CA	IND	9 ALS No FD response	60,000 calls/year	Paramedics	Screen calls by algorithm. Prioritize. Refuse 25% of calls.
Chicago, IL	FD	46 ALS	365,000 calls/year 140,000 runs/year	Some EMTs Some not	Screen calls by flip chart. No prioritizing. Refuse 35% of calls.
Miami, FL	FD	7 ALS FD as 1st responder	28,000 calls/year	Some EMTs Some not	Screen calls by "instinct." Refuse 30% of calls.
Denver, CO	IND	7 ALS FD as 1st responder	40,000 calls/year	EMTs and Paramedics	Go on every call. Determine if life-threatening call. Screen by experience.
Cleveland, OH	IND	5 ALS 7 BLS No FD response	66,000 calls/year	EMTs	Go on every call. Prioritize by algorithm.
Washington, DC	FD	4 ALS 13 BLS FD as 1st responders	90,000 calls/year	EMTs	Go on every call in system. Determine priority and response by category.
New York, NY	IND	34 ALS 104 BLS No FD response	700,000 calls/year	EMTs and Paramedics	Prioritize calls. Screen calls by algorithm. Screen out 25% of calls.

Figure 8.8 Many localities that handle a large number of emergency medical calls have instituted a system for screening calls to cut down on system abuse. *Reprinted from the May, 1983 issue of Firehouse Magazine, 515 Madison Ave., New York, NY 10022.*

COMPUTER-AIDED DISPATCH (CAD)

In some cities computers, not people, perform many dispatch functions. Many departments have found that CAD can significantly shorten response time or enable dispatchers to handle a greater volume of calls. CAD can also reduce the amount of voice communications between fire alarm dispatchers and responding units. A CAD system can be as simple as one that retrieves running card information or as complex as one that selects and dispatches units, determines the quickest route to the scene of an emergency, monitors the status of units, and transmits additional information via mobile data terminals.

Most fire departments that have computers use them to retrieve response information. Dispatchers can enter the address of an alarm and receive a list of units predesignated to respond and such information as quickest routes to the scene of the alarm, hydrant locations, special hazards, and bridge and street closings (Figure 8.9). For example, in Minneapolis, "telecommunicators"

TYPICAL DATA DISPLAY

```
16:10     10/12/78     100 INDEPENDENCE DR     RESCUE

          INITIAL ALARM              2 ALARM WAITING

BEST ROUTE: SOUTH ON LIBERTY, RIGHT ON WASHINGTON BLVD 3 BLOCKS,
            LEFT ON INDEPENDENCE DR

CROSS STREETS: WASHINGTON BLVD AND CAPITAL ST

HYDRANTS AT #98-120-190

INVALID BEDRIDDEN ON SECOND FLOOR
STAIRS TO SECOND FLOOR IN FRONT AND REAR OF STRUCTURE
TEL # OF INVALID'S DOCTOR 465-4486

CAUTION: OWNER STORES FUEL OIL IN BASEMENT

                    MAP GRID G7
DISPATCH    1ST ALM   LDR 1    PMP 4    PMP 7    CHF 11
            2ND ALM   LDR 3    PMP 2    PMP 9    CHF 12
            3RD ALM   SNK 41   PMP 8    PMP 5    CHF 15
```

Figure 8.9 With computer-aided dispatch, such information as best route to the scene of an alarm, hydrant locations, and storage of hazardous materials can be retrieved and displayed rapidly. *Courtesy of Statustronics Corporation.*

take calls for fire, police, or ambulance and use a cathode ray tube (CRT) to enter information about the incident into the computer. A specific fire, police, or ambulance dispatcher then takes control of the incident and sends the proper unit on the way. Fire dispatchers can also get fire inspection and running card information on their terminals and contact the proper units by telephone or radio.

In more complex CADs the computer designates the appropriate units to be dispatched. In Phoenix, the computer system is programmed with the distance from each fire station to each geographic location in the city. When selecting units to respond to an incident, the computer determines the capabilities required and finds the closest units with those capabilities at the moment. If the apparatus is in the station, the alarm is received on the station terminal; if it is in the field, the alarm is received on the mobile data terminal.

Some computers have been programmed to switch signals from the central dispatching unit to the proper location. This feature has been used to send signals received at the alarm center to the station housing the responding units, to switch on lights, and to open apparatus bay doors.

Special pre-fire planning programs that interface with the central computer can provide invaluable information to units as they respond to an alarm. The Phoenix, Arizona, computer automatically transmits information to the mobile data terminals at the time of dispatch. This includes responding units, map coordinates, and assigned radio channel. If there is an occupany file for the address of the incident, such data as building contents, floor plans, hazardous materials, and hydrant locations are automatically transmitted.

In some systems, dispatchers will key in status changes as they are received, and the computer notes them on a status board. Systems with mobile data terminals (MDTs) have status buttons that transmit routine messages directly to the computer (Figure 8.10). The Phoenix MDTs also have a full keyboard that allows additional communication with the computer or any other terminal in the network. The officer in command of an incident can transmit instructions to other units on the call or instruct the system to dispatch or recall units. All of these messages can be sent without voice contact.

New York City has the Automatic Information Dispatching System (AIDS) that provides information on 10,000 buildings. These have undergone structural alterations that would affect fire fighting operations such as reinforcements or false roofs. In Huntington Beach, California, fire and police keep up to 12 different types of data for each occupancy; these data are automatically printed out at the time of dispatch.

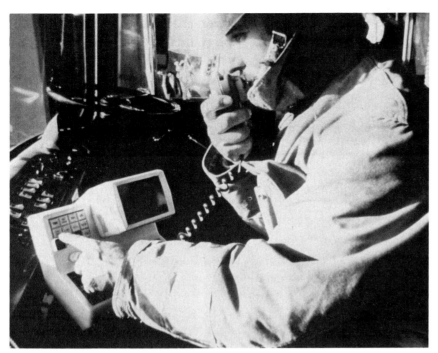

Figure 8.10 In systems that have mobile data terminals, status changes can be sent directly to the computer without voice contact. *Courtesy of Phoenix, Arizona, Fire Department.*

Two Bell System options, Automatic Number Identification (ANI) and Automatic Location Identification (ALI), can make dispatching time even shorter and cut down on false alarms. Costa Mesa, California, which shares fire and police dispatching, has subscribed to ANI and ALI. Coupled with the city's CAD-911 system, the address of the calling party is automatically displayed on the screen when an emergency call comes in. The dispatcher has only to verify the address and enter the appropriate incident code to complete the dispatch.

New York City spent $15 million for its STARFIRE system. It has 16,000 alarm boxes tied to 500 terminals, 12 microcomputer systems, and 14 computers. Alarms come in to small computers in borough communications offices that send them in to a central computer for a recommendation on which units should respond. The recommendation goes to the borough communications officer, where a dispatcher has to approve it before it is sent on to individual fire stations. From alarm receipt to dispatch takes an average of 40 seconds.

Not all fire departments will need or be able to afford computers with all these capabilities. Fire Chief Raymond C. Picard of Huntington Beach, California, notes that putting together the data base required for computer-aided dispatch can take a lot of time and money. He suggests that chief officers consider two key issues in deciding whether to make the expenditures for their departments. One issue concerns dispatching load or frequency. Chief Picard says that a department that typically has only one emergency at a time can do without a computer. Departments that might have more activity, or that may need to transmit hazardous materials information could use a computer to advantage.

Even a small computer may be helpful. The Fitchburg, Wisconsin, Volunteer Fire Department, on the other hand, bought a $2,800 computer from a commercial outlet and used it to store pre-fire planning information available to dispatchers as soon as they code in an incident address. Responding units get the information by radio.

The other issue involves how much special information fireground commanders have to have. The computer can handle a lot of information, but maintaining it can cost a lot. Chief officers have to decide if the benefits can justify the expense. Chief Picard suggests that chief officers who know little about computer technology put together a team to determine what the department needs and how a computer could help the department meet these needs, if at all. The team might want to call in outside consultants or use local government employees to help with the task. Staffers can visit other departments that use computers to see what systems can do, and how they are set up. Also pay particular attention to the different ways to acquire computer operations and consider the advantages and disadvantages of each. Several people have advised that departments select software vendors who specialize in designing emergency systems.

Basically, there are three ways to get a computer: purchase, lease, or lease-purchase. Direct purchase allows a department to get a computer with a one-time expenditure, although funds will have to be spent later for maintenance, repair, and updating. Leasing allows the department to have the same capabilities for smaller expenditures at regular intervals, and the computer firm will usually maintain and repair the computer as part of the lease agreement. The department itself will have to handle any upgrading that needs to be done with improvements in technology. Lease-purchase agreements might be the best of both worlds. The department can work toward eventual ownership and get repair, maintenance, and upgrading while making relatively small periodic payments. If the department decides the system does not meet its needs, it will not be out such a large amount of money.

Some departments subscribe to computer services that set up CRTs at one or more locations and lease signal circuits linking the CRTs to a computer located elsewhere. The service can provide programs, maintenance, and keypunch operators if the department does not want to train its own.

EVALUATING THE COMMUNICATIONS SYSTEM

The communications system is the vital link to the alarm center, fire stations, and the units in the field and on the fireground. There are four areas of communications that must operate smoothly if a fire department is to perform effectively:

- From the public to the alarm center
- From the alarm center to the fire station or mobile unit (dispatching)
- On the fireground, to and from central dispatching and other units
- Interdepartmental (nonemergency) communications

Whether the system is simple or complex, the same general areas must be evaluated when determining whether the system is meeting the needs of the department and the public it serves:

- Integrity of the alarm center. It should be in a secure area safe from sabotage and natural elements.
- Reliability of the communications center. What backups have been provided in case the main system malfunctions? Is the system self-diagnostic?
- Speed at which dispatching can be accomplished. The department must be able to dispatch apparatus and ambulances within a certain period of time to do its job effectively.
- Ability to contact support and combat forces quickly and effectively.
- Integration with other emergency services.
- Ability to handle major emergencies (mutual aid, recall of personnel, contact with special rescue or emergency teams).
- Quality and reliability of communications on the fireground. Can incident commanders get information quickly for evaluation and analysis?
- Sufficient radio channels to handle communications during peak periods as well as administrative and nonemergency radio traffic.
- Ability of emergency medical personnel to communicate with physicians at base hospitals.
- Cost effectiveness and maintenance of system.
- Ability of communications system to be expanded or updated.
- Adequacy of dispatcher training. Dispatchers must be thoroughly familiar with fire dispatching and emergency medical dispatching procedures.
- Number of personnel needed to operate system efficiently.

Before attempting any changes a study should be performed of the community's needs and the department's ability to handle those needs. Determine what parts of the system need improving or replacing and what the costs will be. It may be more cost-effective to install new systems that are initially more expensive but offer greater reliability, speed, and the possibility for expansion at a later date. On the other hand, bigger is not necessarily better. A smaller department does not need computer-aided dispatch but may find that a personal-sized minicomputer addresses its information storage needs perfectly. Consult with other departments that have similar needs to see how their systems are working and to learn what changes they would make. Radio and computer technology are changing so rapidly that expert advice is needed to learn what options are available. Proceed with care: contracts with vendors of communications systems should include guarantees of performance to ensure reliability.

Safety in the Fire Service

9

CHAPTER 9 OBJECTIVES

1. Analyze injury statistics and accident causes for firefighters, and review standard operating procedures and training techniques for safety; submit proposals for improvement in training programs.

2. Monitor operations on the fireground for adherence to safe operating procedures.

3. Establish and maintain a program designed to improve cardiovascular fitness of firefighters.

Chapter 9
Safety in the Fire Service

The problem with fire service safety is demonstrated by the fact that the fire service has the highest rate of occupational injury and death in the United States. Why the fire service has not instituted satisfactory programs to reduce firefighter injuries is unclear. What is clear is that fire service administrators must now shoulder the responsibility for safety.

Safety is a matter of management commitment and attitude changes. The responsibility and obligation for safety lies with management, as H. W. Heinrich first pointed out in the 1930s when he introduced the "domino theory" of accident prevention. The theory describes an accident as a row of dominoes set so that when one falls it knocks over the next. Heinrich's dominoes are

- Lack of management control
- Attitudes and background
- Unsafe acts and conditions
- The incident
- The actual loss

He theorized that accident loss can be prevented by removing any one of the dominoes (Figure 9.1 on next page). Heinrich added that management, which pays for the loss, is in the best position to control these factors; therefore, management must be committed to accident prevention.

No officer wants a firefighter hurt or killed. Then why does the present level of injury and death continue? Many accept the fatalistic theory that accidents are inseparable from fire fighting. The truth is that they probably are, at least until the fire service changes some of its basic attitudes about safety.

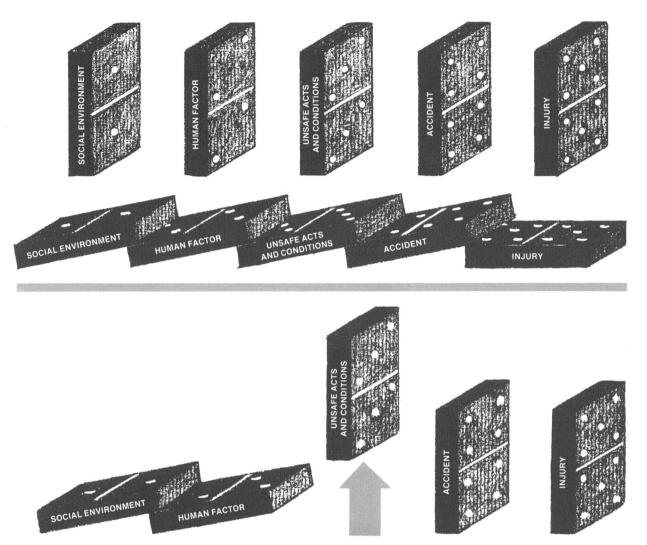

Figure 9.1 Many accidents are caused by a combination of inattention to safety procedures, poor attitude, and lack of supervision. Changing one factor can break the chain of injuries.

The fire department has to take a positive stand that losses from injury are not synonymous with fire fighting. Fire department administration has to accept its obligation to reduce the number of injuries and deaths, realizing that this will require a change in the department's attitude toward safety. There are some basic fire service misconceptions that must be eliminated before the fireground will become safer:

- Injury and death are accepted occupational risks in the fire service.
- Firefighters cannot control their work environment.
- It is heroic to get hurt.
- Safety is a humanitarian issue.

While fire fighting is inherently dangerous, firefighter injuries and death do not have to be inevitable. The fire department has to make a choice between heroics and professionalism. Fire professionalism requires firefighters to work in dangerous environments, but they can reduce the risk with the proper mental attitude, physical conditioning, and protective equipment.

Any change in attitude must begin with top management. If the chief officer does not believe a safety program will reduce losses, it probably will not. The officer should enforce a directive making it clear that the department will not tolerate unsafe working habits (Figure 9.2). Such a statement will go a long way to dispel the attitude that getting hurt or killed on the fireground is heroic. This "macho" concept must be replaced by the idea that injury and death on the fireground are unnecessary wastes of fire department resources. It is the mark of professionalism to train and equip firefighters to function on the fireground safely. The cost of management support and safety training is high, and the task begins with convincing the jurisdiction to spend the money.

SAFETY AS AN ECONOMIC ISSUE

Realizing that safety is not a matter of fate is a good start, but another major hurdle to overcome is obtaining funds. Histori-

Figure 9.2 The attitudes and actions of the chief set the tone for the whole department. The chief must make it clear that unsafe working habits will not be tolerated.

cally, fire departments have had a hard time obtaining funds for safety, and it will be even harder in the future. Administrators who control the funds will not be impressed with humanitarian appeals, especially when it comes to safety; the department must learn to talk in terms comptrollers understand. Firefighter safety must be approached as an economic, not a humanitarian, issue. Administrators must argue that it costs more to allow unsafe work habits than it does to fund a program to reduce losses.

First, show the costs of operation under the present safety program by listing the direct and indirect costs incurred from injury and accident loss. Direct costs include medical expenses, personnel and equipment replacement, and other money paid out in immediate response to an accident. Indirect losses are less obvious. The insurance industry has studied the relationship of direct to indirect accident costs and developed what is known as the "iceberg theory" of accident costs (Figure 9.3). The theory says that for every dollar of direct costs, there are four dollars of indirect costs, which fall into four broad categories:

- Loss of individual expertise
- Loss of unit efficiency
- The high costs of workmen's compensation
- The cost of processing injury claims

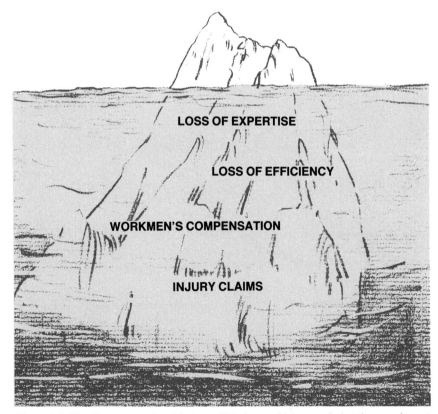

Figure 9.3 The direct costs of accidents and injuries are high enough, but they are far outweighed by the indirect costs.

Documenting these indirect costs is not easy, and requires that the department spend extra effort in the early stages of program development.

DOCUMENTATION

Documenting costs begins with an effective accident reporting system, which is essential to safety program administration. Accident reporting is the foundation because it provides the reference for economic focus. Administrators cannot show the real losses if the reporting system does not reflect the accidents that occur. Moreover, administrators must know the type of injuries firefighters suffer in order to take corrective action. With a good record system, the fire department administration can document the need for economic support, administer the program, and show firefighters its relevance (Figure 9.4).

Personnel Board of Jefferson County
Employee Relations and Safety Branch
ERS-1

INJURY REPORT
FIRST REPORT OF INJURY

To be filled out by immediate supervisor and forward original copy to the Personnel Board within twelve (12) hours of accident

NAME OF EMPLOYEE _____ CLASSIFICATION _____

JURISDICTION _____ DEPARTMENT _____

EMPLOYEE'S AGE _____ DATE OF INJURY _____ TIME _____ a.m. / p.m.

LOST TIME ACCIDENT   Yes _____ No _____

DETAILS OF ACCIDENT
(This information is for use in preventing similar accidents. Please answer all questions.)

1. What task was employee performing? _____

2. How was employee injured? _____

3. What did employee do unsafely? _____

4. What equipment was defective? _____

5. What steps should be taken to prevent similar injuries? _____

6. Was accident reported immediately? Yes _____ No _____. If No, Explain _____

7. Did employee require medical attention as a result of this injury? Yes _____ No _____.
   If yes, give name and address of doctor and/or hospital _____

Figure 9.4 Complete records must be kept of accidents in order to compute costs and to determine what corrective measures are needed.

The fire department administration can also use records to show that the safety program is producing a reasonable rate of return on the investment if administrators can develop a concept of accident-to-cost relationships. The insurance industry, with its experience in business and industry, has come up with a "trivial many and vital few" concept. The concept indicates that 25 percent of all accidents, the "vital few," account for 75 percent of the losses. If the fire department could control these accidents, it could reduce losses by 75 percent. This is a substantial investment return. On the other hand, if 75 percent of the accidents account for only 25 percent of the costs, it is economic folly to devise expensive safety programs to prevent them. The insurance industry found out that the "trivial many" are best controlled by inexpensive administrative policies and procedures. If administrators fail to act on the concept, their programs will concentrate on the trivial many while ignoring the vital few. Smart safety program management will institute administrative policies and procedures to control the minor accidents and invest safety program money to prevent accidents.

The insurance industry also discovered that incidence rates are directly related to employee satisfaction and management-labor relations. Therefore, prudent safety program managers will incorporate employee relations into administrative control to reduce on-the-job accidents. A key element is a department safety committee, which represents the largest investment in an accident safety program. The committee should provide liaison between labor and the department management; however, administrators should not let the committee become a battleground for political interests and labor negotiations.

If the committee is given too much authority, it can become a hazard instead of an asset. To be effective, the committee must understand the scope of the safety program, which will be defined locally. When safety committees are given decision-making authority, they can undercut the prerogatives of the chief officer. Chief officers should limit the responsibilities of the safety committee to identifying hazards and developing suggestions to correct the problems.

On the other hand, chief officers should follow up on committee suggestions. In all cases, chief officers should report how they acted on committee suggestions or explain why they cannot be implemented. If there is no feedback, committee members will quickly come to view the safety committee as a token management effort. Then the committee can become a source of strife between management and employees.

There are no rules for forming a safety committee, but administrators should consider bringing together representatives of both labor and management. The leadership should be decided by the committee members themselves. Administrators should also make it clear that the committee is a management tool for accident control, not a labor-negotiated right unless it is so stipulated in the contract.

The department should also consider appointing a safety officer who would report directly to the chief officer. This person might even be a member of the safety committee, possibly the secretary. Be sure to confine the officer's duties to education and administration rather than discipline. The safety officer should not take on the role of a police officer or issue citations for safety violations. Firefighters must be made to understand that ultimately they are responsible for their own safety.

PHYSICAL FITNESS AS A SAFETY ISSUE

Current death and pension statistics strongly indicate that improving firefighters' physical fitness can be worth the money. For example, nearly half of all on-duty firefighter deaths are caused by heart attacks, and the cost of rehabilitation after a heart attack is one of the most expensive of all rehabilitation efforts (Figure 9.5). Heart disease in the fire service is one of the loss categories that violates the concept of the trivial many and vital few. The occurrence rate is extremely high and the cost per incident is substantial. Firefighters are prime candidates for heart disease, which typically begins when blood vessels are irritated by toxins entering the bloodstream. Scarring occurs as cholesterol and fatty deposits coat the irritated vessels. Eventually the

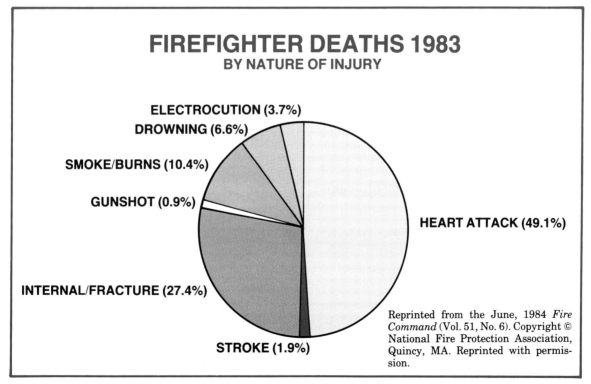

Figure 9.5 Heart attacks are by far the most common cause of firefighter deaths.

vessels' ability to transfer oxygen and food to the muscle tissue is impaired, and increased physical activity will further damage muscle tissue. Often the heart muscle itself is damaged, reducing the heart's ability to support the body. After a time, the heart can suffer so much damage that it will fail or become so weak that it can no longer support other critical body organs.

The firefighter's susceptibility begins with an elevated exposure to the toxins that cause the initial irritation. The two most common irritants are carbon monoxide and adrenaline. In addition, there are many other extremely toxic substances in smoke and fire gases. A large percentage of firefighters also smoke cigarettes, which introduces CO and other toxic substances into the bloodstream. Because so many firefighters smoke in the fire station, many of the nonsmokers are subjected to the same cardiovascular irritants.

The effects of fireground exposure and cigarette smoking are compounded by the excitement of the job, which causes high levels of adrenaline in many firefighters. A Boston study showed that some firefighters' adrenaline levels rose 1½ times during alarm response. Some suspect that the prolonged exposure to a combination of toxins has a combined effect far greater than the effect of each toxin alone.

The firefighter's plight is worse because of a false sense of physical fitness. Firefighters perform strenuous tasks regularly, and most assume that this means they are in good shape. In most stations in the nation, firefighters are fooling themselves. Most firefighters do not respond to enough working alarms to maintain an adequate level of physical fitness simply by doing their jobs. In fact, the average firefighter has a higher level of body fat than the average American male. Overweight people are also more likely to have high blood pressure, which further strains the heart.

A good physical fitness program concentrates on preventing or reducing the effects of these conditions. The most common mistake made in developing a program is concentrating on strengthening muscles instead of conditioning the heart and lungs. Fitness programs should focus on exercises that improve the heart rate and blood flow. A strenuous weight-lifting or exercise program might or might not improve the heart. In fact, an aerobics exercise class might meet the needs of the fire service better than a weight-lifting program. Administrators must be careful that the program they adopt is aimed primarily at improving heart condition and body flexibility, with increased muscle strength as a secondary objective.

In most cases, heart condition can be improved without adversely affecting the individual's health. The conditioning begins with modest exertion and gradually increases. A regular program of this type will gradually strengthen the heart to meet its existing level of exercise. Generally, physical conditioning, muscle tone, and body fat percentage will also improve.

A qualified medical staff should monitor a new fitness program to make sure firefighters do not inadvertently damage weak heart tissue during their first efforts (Figure 9.6). Before instituting the program, the staff must screen firefighters to identify any persons who might not be physically capable of beginning the program. They will have to get special medical attention and therapy to overcome their poor cardiovascular condition. The screening also provides a benchmark against which other firefighters can base their program.

Figure 9.6 Firefighters will benefit most from activities that are designed to improve the heart and lungs.

The program must be tailored to each individual with a goal of slowly increasing the heart rate during moderate exercise until the body becomes conditioned to functioning adequately at elevated heart rates. Medical consultants armed with screening results should decide what increase in heart rate to attempt. Programs are often like a management-by-objectives system: each firefighter, with advice from the medical advisor, sets conditioning goals to be met in a certain amount of time. Since each firefighter is in a different physical condition, the program for each firefighter should be different. As the program continues and firefighters' physical condition improves, administrators might want to begin an organized exercise program to maintain heart fitness, increase physical stamina, and improve respiration. Do not neglect organized recreation. Avoid contact sports such as basketball and football but consider instituting programs in track, swimming, and other noncontact sports. The department could build its own facilities or work out agreements to use other public or private facilities in the community.

Administrators can conduct other phases of the safety program at the same time. For example, a breathing apparatus program should reduce exposure to the products of combustion. The program should go beyond supplying breathing apparatus and training firefighters how to use them. It should also include firm administrative controls that require firefighters to wear breathing apparatus on the fireground whether inside or outside the structure (Figure 9.7). Firefighters have a terrible misconception that only firefighters inside the structure need breathing apparatus, but in fact any person on the fireground is in danger from toxic gases. When smoke shifts, few firefighters will return to the apparatus for breathing air. If firefighters are already wearing breathing apparatus, they can easily connect the air supply.

The fire administrator who proposes and enforces such a policy will face a lot of initial opposition. Firefighters will complain that breathing apparatus are too heavy to be worn all the time. Budget officials will complain that it is too expensive to supply firefighters and officers with constant air. The logistics officer will complain that a lot of air is used unnecessarily. The administrative officer has to demonstrate that the long-term savings far outweigh these other considerations.

Some departments are now refusing to hire smokers. The initial court tests of this practice have favored the fire departments. They have been able to show that

- There is a documented association between cigarette smoking and heart disease.
- Firefighters are exposed to high levels of smoke and fumes on the fireground; these hazards are further complicated by cigarette smoking.
- Heart disease is an occupational hazard in the fire service.

Some courts have held that cigarette smokers constitute an elevated risk to the fire department's pension program because of their smoking.

Some fire departments have also instituted dietary programs that monitor each firefighter's weight. Administrators have established weight limits and firefighters above the limits have to reduce. The programs also provide diet information to firefighters and their families. Some departments have even employed dietitians to counsel firefighters on their eating habits and to develop diet plans for station meals.

SAFETY PROCEDURES
SELF-CONTAINED BREATHING APPARATUS

PHOENIX FIRE DEPARTMENT
STANDARD OPERATING
PROCEDURES
M.P. 205.01
3/1/78 Rev. Page 1 of 2

It is the policy of the Phoenix Fire Department that all personnel expected or likely to respond to, and function in, areas of atmospheric contamination, shall be equipped with, and trained in, the proper use and maintenance of the self-contained breathing apparatus (S.C.B.A).

Each member of the Operations Division shall be accountable for one (1) S.C.B.A. and shall check that S.C.B.A. for condition at the beginning of each shift and after each use, or at any other time it may be necessary to render the equipment in a ready state of condition.

Company officers shall assign a specific S.C.B.A. to each member of the crew. Each crew member will be responsible for the proper use and function of that S.C.B.A. If an S.C.B.A. is found to be functioning improperly, it shall be taken out of service, red tagged, reported, and replaced as soon as possible. Replacement S.C.B.A.s may be acquired through Support Services.

All personnel shall utilize the provided S.C.B.A. when encountering the following emergencies:
- Above ground level.
- Below ground level.
- Contaminated atmosphere.
- Situations where it is likely that the atmosphere may be contaminated.

Resist the tendency to prematurely remove breathing apparatus during routine fire situations. We all must be aware of the respiratory hazards which exist in ordinary as well as the extraordinary fire situation. It is generally true that *carbon monoxide levels increase during overhaul*, due to the incomplete combustion of smoldering materials.

Do not remove your S.C.B.A. until the atmosphere has been determined to be safe to operate within. *Either use your S.C.B.A. or change the atmosphere.*

The determination as to removal of breathing apparatus will be made by company or sector officers in routine situations. In complex situations, particularly when toxic materials are involved, the Safety Officer and/or the Fire Protection Engineer should be consulted on this decision.

Figure 9.7 A policy on wearing self-contained breathing apparatus, as well as strict enforcement, will demonstrate management's commitment to safety.

The Political Arena

10

NFPA STANDARD 1021
STANDARD FOR FIRE OFFICER
PROFESSIONAL QUALIFICATIONS

Fire Officer V

6-1 Political Science.

6-1.1 The Fire Officer V shall demonstrate knowledge of federal, state, and local legislation affecting fire protection in the authority having jurisdiction.

6-1.3 The Fire Officer V shall demonstrate knowledge of the lawmaking process in the state government of the jurisdiction having authority.

6-1.4 The Fire Officer V shall demonstate knowledge of the lawmaking process in the jurisdiction having authority.

6-1.5 The Fire Officer V, given an actual or simulated situation requiring new legislation, shall demonstrate the ability to draft the proposed legislation.

6-2.6 The Fire Officer V, given a summary of current or pending legislation of interest to the fire service, shall identify legislative intent and inadequacies in new or current laws.

6-3 State and Local Government.

6-3.1 The Fire Officer V shall demonstrate knowledge of authority of the branches of government in the jurisdiction having authority.

6-3.2 The Fire Officer V shall identify the cabinet post and/or departments of government in the jurisdiction having authority.*

Fire Officer VI

7-4.3 The Fire Officer VI shall demonstrate the ability to interpret fire service legislation at the local, state and federal level.*

The above NFPA standards are addressed from a general management perspective.

*Reprinted with permission from NFPA No. 1021, *Standard for Fire Officer Professional Qualifications*. Copyright 1983, National Fire Protection Association, Boston, MA.

Chapter 10
The Political Arena

The primary functions of state and local governments are to allocate public resources and to set public policy. The goal of groups or individuals who engage in politics is to guide or influence governmental policy. Traditionally, the fire service has stayed out of the political arena, except to lobby for legislation related to labor relations, job benefits, and job security.

Until recent years, fire departments were fairly sure of receiving requested funds, and there was little need to prove that those funds were being used in a cost-effective manner. Times have changed, though, and the fire service is paying for its lack of foresight and inattention to political matters. In recent years, taxpayer revolts and economic recessions have left state and local officials with the task of dividing a smaller pie. The public will no longer support public safety regardless of cost, and city managers are keenly aware that citizens expect them to see that their tax dollars are spent wisely.

Not only must fire departments operate with reduced budgets and fewer personnel, they must also compete with other departments for funds and be prepared to back up their requests with solid facts. Many officers are trained only in suppression techniques; consequently, they lack the managerial and administrative skills necessary to deal with city managers, labor representatives, and business officials.

To compete successfully at an increasingly technical and sophisticated level, the chief must be a skilled manager and do a good job of selling the department and its services to the community. A department that is professional, active, and highly visible will achieve a strong level of local support. The ability to apply this type of positive political pressure will make city managers reluctant to cut a service that clearly benefits the community. A chief whose department is well run and who maintains ongoing, cordial relations with city leaders will find the political climate favorable for department requests and programs.

BECOMING AN EFFECTIVE LEADER ON THE LOCAL SCENE

The chief is part of the city's management team, but is also competing with other departments for public funds. To be an effective manager the chief must be well trained in management and administration, budgeting techniques, and be familiar with the community and its needs. To be an effective competitor for funds, the chief must be prepared to demonstrate exactly why the department should receive the requested funds, backing up statements with solid data. The political savvy required to achieve favorable outcomes in both areas rests on preparation and good communications.

Firefighters preplan for incidents, taking into account everything that can impede a successful attack. Preparing for budget hearings, gathering support for programs, or trying to pass fire safety legislation requires the same kind of homework to be successful.

The Chief as Manager

What kinds of expertise does the chief need to function effectively as a manager? To be most effective, the chief needs to be

- *Well informed about the community*. The chief should be a generalist who is knowledgeable about all facets of the community. As a manager, the chief should be familiar with demographics, resource needs, statistics, economics, land use studies, and environmental impact statements.

- *Capable of making intelligent presentations based on efficiency and cost effectiveness*. The fire department is a service business supported by taxpayers' dollars and must be treated as such. City managers want to be able to equate dollars spent with services delivered. They need a detailed plan of action, based on management by objectives, that clearly states where the department is, where it is going, and the amount of money that needs to be spent in each area of the department. The chief must be able to present professional alternatives that detail the services that can be delivered at each level of funding (Figure 10.1).

- *A team player*. The chief is part of the city's management team and is expected to be able to work effectively with high-level city officials and their staffs as well as with other department heads. A chief with a reputation for hostility or defensiveness will be neutralized by other department managers, who are also competing for tax dollars.

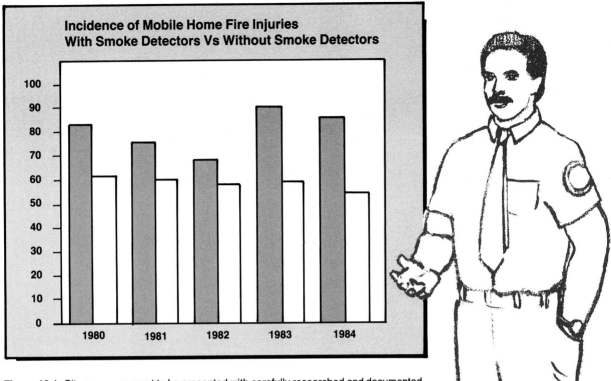

Figure 10.1 City managers want to be presented with carefully researched and documented plans that detail the fire department's present and future needs.

- *Futures oriented.* The department's plans must extend beyond the next 12 months. The chief must be able to present long-range plans and objectives that are designed to improve service capabilities and meet the community's changing needs. What services will the fire department be delivering five years from now? What problems will the department be facing five years from now? What groundwork must be laid to ensure that the department (and the city) reaches those goals?

- *Familiar with the techniques and mechanics of budgeting.* These include planning, programming, and evaluation. The budgeting process is becoming highly technical, and the chief who fails to understand its use as a managerial tool will see budget decisions made by others.

Increasing the Visibility of the Department

Even if a department is well run and performs services efficiently and in a cost-effective manner, it will not count for much if no one knows about it. Says one chief: "Any public exposure is better than what we've had in the past." Although the fire department provides a vital service, so do other public service departments (police, public works, schools), and the chief must learn to work closely with other departments and city managers and their staffs on a year-round basis. The fire department does have some advantages: its services are vital to public safety, its work (particularly emergency medical care) provides visibility, and stations indicate an established presence in the community. The task is to capitalize on these advantages to gain and keep public support.

- Before attempting to increase the department's visibility, make sure that the department is unified internally. Establish clear and regular communications with department personnel and union representatives. Everyone in the department should know what activities the department is engaged in and for what purpose. Involve all personnel in participatory management so that personnel will not just feel like employees, but like part of an organization with a clear purpose. Explain programs to union representatives so that parts of the department do not start working at cross-purposes to one another.

- Work on making your department something to be proud of. A department known for its immaculate, well-maintained stations, friendly and professional crew, and interest in the community will have an easier time justifying its requests.

- Do not wait for the community to come to the fire department, get the fire department out to the community. The opportunities for local exposure are many: speak before community groups such as the Chamber of Commerce and the League of Women Voters; take public education programs to schools, group homes, and churches; perform regular inspections of hydrants, homes, and businesses; teach CPR classes both in the station and out in the community (Figure 10.2). Every time a member of the department in-

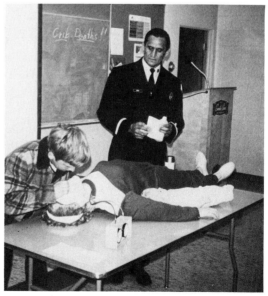

Figure 10.2 The fire service can demonstrate its commitment to life safety in many ways. Here citizens are being taught to perform CPR.

teracts with someone in the community, there is the potential for creating goodwill. Be sure that those who deliver programs are evaluated for effectiveness.

- Hold regular social events to which the public is invited. Be sure to invite members of the city council and legislators from your district. Invite the same people to training sessions and firefighters' board meetings.

- Get the media on your side. Invite the local newspaper in to talk or ask about writing columns on fire safety for the paper. The media can be an excellent way to perform some public education and earn some free publicity for the department (Figure 10.3).

PUBLIC SERVICE ANNOUNCEMENT

SUBJECT: Fire Prevention
FROM: Your Fire Department

(50-Second Announcement — 100 words)

Cold weather is coming, and a fire's going to feel good before long! But it wouldn't feel good if that fire were burning down your home . . . and if you don't take precautions, that can happen! So have your furnace, chimneys and flues inspected and cleaned if necessary before the time comes to use them. See that any heaters you own are clean and in good repair before you turn them on. Provide a screen for your fireplace, and see that it is always in place whenever you light an open fire. Make sure that all the fires that burn in your home are *friendly* fires!

Figure 10.3 Establishing good relations with the local media enables the fire department to perform valuable public education.

- Take an active role in community affairs. Have departmental representation at zoning meetings and public hearings. Attend political and community functions. Vote, and encourage members of the department to vote, but do not involve the department in partisan politics. During elections, take the opportunity to buttonhole candidates in a friendly way and state the case for the fire department.

All of the above are methods of generating positive political pressure for the fire department, but it is obvious that friendly relations and open communication lines must be maintained on a year-round basis. Do not call on city officials only when you want something. If the department wants community support for its programs, it must be willing to support other departments' programs and the community at large.

Working to Pass Local Legislation

Sooner or later, the fire department's strong and active presence in the community will extend to the area of legislation. Working to pass ordinances covering such safety issues as automatic sprinklers, smoke detectors, fireworks, and transportation of hazardous materials is the logical extension of the fire service's commitment to safety. Whether the proposed ordinance would update an existing code or adopt a new one, the steps needed to conduct a successful campaign are the same: preparation, education, and alliance building.

Thorough preparation and a careful analysis of the effects of the proposed legislation are absolutely necessary. Before talking to anyone in the city government, the fire department must gather reliable, expert knowledge. The chief must be prepared to answer all questions that are likely to arise concerning the effects of the proposed ordinance. Exactly what is the problem and what would the ordinance do to solve it? What will the cost be to the city or to those it affects? Contact other fire departments that have dealt with similar problems and obtain copies of their ordinances. Ask questions. What were the keys to their successful campaigns? What did the ordinance accomplish? What groups were opposed to the ordinance and why? What penalties were assessed for failure to comply with the law? What would they do differently if they had it to do over again?

Since council members do not like surprises, the next step is to schedule meetings with the city manager and city attorney. Outline the problem in terms of the fire and liability danger it presents to the community and the ability of the fire department to deal with it.

Describe how the city and its citizens will benefit from the proposed legislation. Point out just what the city stands to gain or lose by failing to adopt the ordinance (loss of life, loss of property, loss of tax revenues). Do not exaggerate or misrepresent the situation: the chief's purpose is to present alternatives for dealing with the problem. The advice and cooperation of the city attorney is needed in order to draft an ordinance that addresses the needs of a particular locality and that will stand the test of law.

The chief should know all members of the city council, their interests, and how they are likely to react to a proposed ordinance. Be prepared to communicate with the opposition. Talk with those who are opposed to the ordinance in order to understand their point of view and to learn what in the ordinance has generated opposition. Stay receptive to questions from the media so that you can perform some public education or correct misconceptions.

The presentation before a city council should be a reiteration of the dialogue that has gone on before. State the fire department's case clearly and succinctly. Use visuals for further clarification if necessary. Keep it brief: the council will have many items on the agenda so do not attempt to monopolize the spotlight or carry on a lengthy discussion.

If the ordinance is rejected, take it seriously, but not personally. Find out what caused it to be defeated and take it into account when making the next attempt. It may take years to lay the groundwork for a proposal that will pass in a few minutes during a council meeting.

STATE POLITICS

Frequently laws governing safety codes, emergency medical regulations, and working conditions for career firefighters are made at the state level. To be effective at the state level requires careful organization, alliance building, knowledge of the legislative process, and the establishment of good relationships with legislators and their staffs.

There are no overnight successes: This process will take a minimum of three to five years. However, the chief who is willing to spend the time will have the opportunity to do the following:

- Become part of the information network on fire and safety legislation in the state.
- Form alliances with like-minded fire, industrial, and citizens' groups throughout the state.
- Help educate the public, press, and legislators about fire safety issues.

The Legislative Process

The political process is a little different in every state, but the passage of legislation follows some basic steps in nearly all states (Figure 10.4 on next page). It is a lengthy process by design: the many steps allow for careful consideration and input from many people. Take the time to learn how the legislative process works in your state. Contact representatives of the state firefighters' organizations or visit the offices of your legislators to learn the specific steps needed for introducing and monitoring legislation.

Learn who the power brokers are: what are their interests, where do they stand with regard to fire safety, and how are they likely to vote?

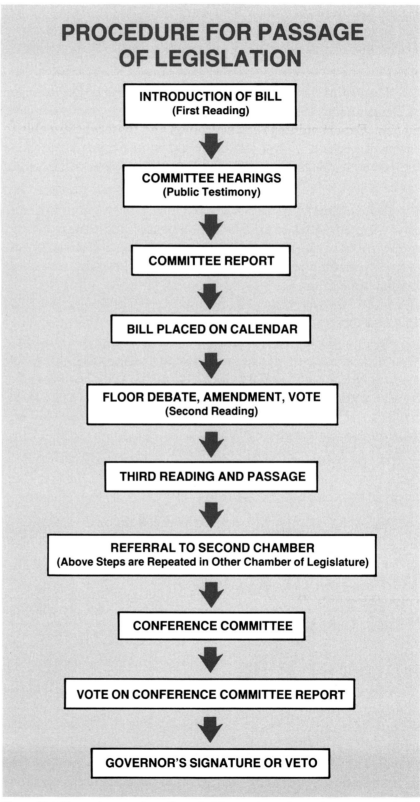

Figure 10.4 Years of preparation, alliance building, and attention to the many steps in the political process are necessary to achieve success in passing legislation.

It is very important to establish friendly relationships with state legislators and their staffs. Do not insist on speaking only to the senator or representative: their legislative and administrative aides prepare position papers and are key people to cultivate. Invite them to workshops and training sessions so they can get a better understanding of the fire service. Send legislators and their staffers information you think they would be interested in; visit their offices and exchange information in a friendly way. Legislators are expected to vote and hold opinions about a large number of issues, but have limited resources for gathering information. Expert opinions are welcomed and those who are able to provide facts and figures have the opportunity to engage in a regular exchange of information.

The Importance of Unity

The necessity for presenting a united front when working to pass legislation cannot be overstressed. If firefighters want their voices to be heard, they must speak with one voice. If legislators hear different opinions from their state's fire service, they have no way of knowing just who is speaking for the fire service. They are more likely to be influenced by a group that is well organized and presents a single point of view. Fire service organizations must learn to work for each other and for the common goal of life safety. Much time will be needed to form alliances with citizens' groups, business groups, and manufacturers in order to pass legislation. Remember, though, that those who are opposed to legislation designed to reduce fire hazards have a vested interest in the status quo. They will be well organized and well funded. Firefighters work together on the fireground: they must do the same in the political arena.

Communications Regarding Legislation

Approach legislation in a very straightforward manner. State your position on proposed bills and be honest about what you think the bill can and cannot do. Have the facts to back up your statements. Do not communicate only when you are opposed to a measure; maintaining a relationship requires support as well. You will not make or keep friends by only being negative. In state politics, as with the local city council, cordial relationships are essential.

Letter Writing

The best time to write a letter is before the bill has been set for a committee hearing, because the number of legislators you will need to contact is far fewer. A well-written, carefully stated position by a respected group of people or on official stationery may have more impact than many letters from people unknown to the legislators. A handwritten letter can also be very effective, especially if the legislator knows the author. Whether you are writing as an individual or as part of a letter-writing campaign, it is important to use your own words. Many letters that all sound alike will have less effect than one that is individually composed.

The letter should be brief and should mention the number of the bill and the author. If the letter is too long, it might not be read by the people for whom it is intended. Send copies to your state association or national groups that you know have an interest in the bill.

If you support the measure, say so and offer your assistance. If you are opposed, state the reasons why and offer an alternative if possible. *Never* get personal and *never* make threats. Just because you support or oppose a bill does not mean the representative will vote your way. Your goal is to establish credibility so that your views will be taken seriously. Threats to set up picket lines or organize against a representative's re-election are unprofessional and will only make enemies.

Follow up on your communications after the bill has been passed or defeated. If you are pleased, write or telephone to say thank you for the support. Give the legislator a plaque or fire helmet expressing the appreciation of the fire service (Figure 10.5). If you are displeased, again state your reasons and ask if there are any alternatives. If the bill has been reported out of committee, there is still a chance that the bill can be amended if enough people feel strongly about it.

Figure 10.5 Show appreciation to legislators who have made special efforts to support fire safety issues.

Public Hearings

Public hearings are held prior to making decisions on a bill. They are rather informal and do not necessarily determine how legislators will vote; however, they do serve some useful functions. Legislators have an opportunity to gauge the intensity of feelings on both sides of an issue, and those who testify get to have their "day in court." The chief who testifies and does a good job can reap several benefits:

- becoming known as an expert on fire safety issues
- making closer contacts with legislators, leading to further exchange of information
- gaining extra visibility for the issue if the hearings are publicized

The most important points to remember when testifying are to be concise and well informed. Be sure your testimony has been thoroughly researched; do not try to "snow" the committee members. Since there may well be a large number of people who wish to testify, take no more than five minutes to present your case. Dress neatly in uniform or a suit.

Lobbying

Generally, direct lobbying is best left to professional lobbyists. In most states, one must register in order to lobby. If the state fire chiefs' organization has a full- or part-time lobbyist or a legislative committee, these persons will be able to cover the legislative scene more intensively. The lobbyists keep track of the status of bills, develop a good rapport with legislators, and can inform the state firefighters when is the best time to make their views known. The fire chief will be able to work more effectively by coordinating communications with the lobbyist so that they reach the right people at the right time.

Appendix A

EQUIPMENT AVAILABLE FOR MUTUAL ASSISTANCE – METRIC CONVERSION

FIRE DEPT.	PUMP LPM CAPACITY	63.5mm HOSE AMOUNT & THREAD	76.2MM AND LARGER HOSE AMOUNT & THREAD	TANKERS AMOUNT OF WATER	GROUND LADDER	AERIAL DEVICES	HEAVY STREAM APPLIANCES
LAKE BLUFF FIRE DEPT.	1 - 3785	244m NST	76.2mm - 15.2 m NST 101.6mm - 609.5m Storz	1 - 3785 L	1 - 7.3 m 1 - attic 1 - roof		deluge set
FORT SHERIDAN FIRE DEPT.	1 - 2838 1 - 1892	792.5m NST	42.6m NST		2 - 10.7m 2 - attic 2 - roof		
BARRINGTON FIRE DEPT.	2 - 5677 1 - 946	305m NST	76.2mm - 472m NST	1 - 11,354 L	1 - 4.3m folding 1 - 7.3m 1 - 9.1m 1 - 10.7m 1 - 12.2m 3 - roof	1 - 25.9m aerial ladder	1 - 1892 LPM 1 - 3785 LPM
COUNTRYSIDE FIRE DEPT.	1 - 3785	305m NST	76.2mm - 274.4m NST	1 - 3785 L	1 - 4.3m 1 - 7.3m		1 - aerial water gun mounted on pumper
FOX LAKE FIRE DEPT.	1 - 5677 1 - 3785 1 - 1325	472m NST	76.2mm - 640.0m NST	1 - 14,000 L	1 - 4.9m 3 - 7.3m 1 - 10.7m 1 - 12.2m 2 - roof 2 - attic	1 - 18.9m aerial tower	2 deluge sets
HIGHWOOD FIRE DEPT.	1 - 2839 1 - 3785	792.5m NST	76.2mm - 290m NST		2 - 4.3m 2 - 10.7m 1 - 15.2m	1 - 27.4m aerial platform	1 deluge gun
KNOLLWOOD FIRE DEPT.	2 - 3785 1 - 2460	290m NST	76.2mm - 83m NST 101.6mm - 366m Storz	1 - 3785 L 1 - 5677 L port.	1 - 7.3m		2 - 1892 LPM pipes
LAKE VILLA FIRE DEPT.	1 - 3785 1 - 2839	457m NST		1 - 2650 L 1 - 5677 L port.	2 - 10.7m 2 - roof 2 - attic		1 - 1892 LPM deluge gun mounted on a 4WD truck
NORTH CHICAGO FIRE DEPT.	1 - 3785 1 - 4731	244m NST	76.2mm - 396.3m NST		1 - 4.3m 1 - 7.3m 1 - 10.7m	1 - 25.9m aerial tower	1 - 1892 LPM 1 - 2839 LPM

FOAM EQUIPMENT	PORTABLE PUMPS LPM	LIGHTING EQUIPMENT	SCBA	AMBULANCES	SPECIAL EQUIPMENT
	1 - elec.-946	2 - 3500 wt. gen. 5 - floodlights	6 - 30 min. MSA	1 - rescue squad	Rear tank dump, generator
360 L AFFF 125 lbs. dry foam mix	2 - 1135	1 - 2500 wt. gen.	6 - 30 min. MSA		K-12, Porta-Power, cribbing, explosimeter, gas generator, smoke ejector, brush equip., air crash rescue gear, proximity suits
2 - pickup units 378.5 L foam in 192 L containers	1 - 378.5	1 - 5000 wt. port. gen.	6 - 30 min. Scotts & spare bottles	1 - MICU 2 - patient	Acetylene torch, air bottles, air cascade sys., smoke ejector, air chisel, K-12, Porta-Powers and come-alongs, salvage equip., 2 Hurst tools (24 and 32), low level strainer, siphon device, 254mm rear dump, 4WD brush unit
	2 - 946	1 - 1500 wt. gen.	6 - 30 min. Survivair	1 - 2 patient	Smoke ejector, water vac, salvage covers, Jeep with small pump, boat
151.4 L foam eductor 1 - foam unit	1 - 1135	1 - 4500 wt. port. gen. 2 - 4.9m light towers w/ 6-5000 wt. bulbs on each 10 kw gen.	6 - 30 min. MSA 6 - 30 min. Scotts	2 - rescue squad	2 smoke ejectors, K-12 saw, floating strain, 2 back packs in basket, Manpower Squad and Divers Van, 3 boats, air bank, Hurst tool, hydraulic saw, elec. chisel, elec. winch, air bags, 2 FWD Jeeps w/378.5 L tanks, Stokes basket, 4WD minipumper brush truck
4 - 192 L cans 2 - foam applicators		1 - 3000 wt. gen. 1 - 2000 wt. gen.	4 - 30 min. Scotts	1 - 4 patient	K-12 saw, Porta-Power
high-expansion foam gen. Angus foam inductor Foamaster nozzle 227 LPM 56.8 L AFFF 1892.5 L deluge		1 - 3000 wt. gen. lighting equip.	2 - 30 min. Survivair		
foam eductor 1 - foam proportioner	1 - 946	1 - 6500 wt. gen. 2 - floodlights	3 - 30 min. Scotts & spare bottles	1 - rescue squad	Generator, smoke ejector, Hurst tool, Porta-Power, K-12 saw, air bank, stretcher, 2 Jeeps with 567.7 L tanks
	1 - 1892 1 - 1324	1 - 5000 wt. gen. 2 - floodlights	6 - 30 min. Scotts	1 - 4 patient 1 - rescue squad	Hurst tool, chain saw, Porta-Power, air chisel, K-12 saw

Appendix B

REFERENCES AND ENDNOTES

CHAPTER 1 REFERENCES

Baltz, Duane. "How a Disaster Differs from an Emergency." *Fire Chief Magazine*. January 1982, p. 58.

A Basic Guide for Fire Prevention Control and Master Planning. Washington, D.C.: U.S. Department of Commerce, n.d.

Bierwiler, David G. "Managing the Fire Service in the Future." *Fire Chief Magazine,* January 1984, p. 43.

Casper, Andrew C. and Roger E. Herman. "Emergency Preparedness — Everybody Must Be Involved." *Fire Engineering,* July 1983, p. 39.

Clark, William E. "Improving Your Annual Report." *Fire Chief Magazine,* January 1970, p. 45.

Coleman, Ronald J. "Master Planning and Emergency Management." *Fire Chief Magazine,* May 1981, p. 61.

Cox, Jackie. "Fire Chiefs — New Roles, New Images." *Fire Engineering,* January 1984, p. 14.

Didactic Systems, Inc. *Management in the Fire Service*. Boston: National Fire Protection Association, 1977.

Giuffrida, Louis O. "The Integrated Emergency Management System." *The International Fire Chief,* January 1984, p.15.

Haney, John. "Disaster Exercise: A Learning Experience." *Fire Chief Magazine,* July 1984, p. 30.

Hawkins, Thomas M. "Building Bridges — The Fire Service's Involvement in Emergency Management." *The International Fire Chief,* January 1984, p. 18.

Hildebrand, Michael S. *Disaster Planning Guidelines for Fire Chiefs*. Final Report prepared under FEMA Contract DCPA 01-79-C-0303, July 1980.

Holcomb, Bill. "Master Planning — an Alternative to Reactive and Incremental Decision Making." *The International Fire Chief,* February 1980, p. 12.

International City Management Association. *Managing Fire Services*. Washington, D.C.: International City Management Association, 1979.

Kitchen, Ross A. "Mutual Aid Pays Off." *Fire Chief Magazine,* May 1981, p. 56.

McElfish, Jack K. "Tapping the Private Sector-A Management Alternative." *The International Fire Chief,* May 1984, p. 19.

Newcomb, Robert and Marg Sammons. "When It's Time to Report." *Fire Chief Magazine,* January 1970, p. 41.

Owens, Kenneth L. "Strategic Planning for Future Fire Protection Needs." *Fire Chief Magazine,* June 1984, p. 51.

Peterson, Carl E. "UFIRS — The New Management Tool for Chiefs." *Fire Command,* February 1974, p. 30.

Rule, Charles H. "Is Fire a Social Problem?" *Fire Chief Magazine,* August 1979, p. 89.

Ryland, Harvey G. "You are at Risk." *The International Fire Chief,* January 1984, p. 12.

Ventimiglia, Mike. "Blueprint for Disaster Planning." *Fire Engineering,* July 1984, p. 30.

CHAPTER 2 REFERENCES

Alexander, Charles A. "Psychological Evaluation of Firefighter Candidates." *Fire Chief Magazine,* September 1981, p. 61.

Anderson, Kirk B. and L.R. Gagnon. "Training and Testing — Partnership and a Route to Equal Employment Opportunity." *The International Fire Chief,* July 1982, p. 81.

Development of a Job-Related Physical Performance Examination for Firefighters. Washington, D.C.: U.S. Department of Commerce, 1977.

Fire Department Personnel Management Handbook: Managing the Entry of Women & Minorities. FEMA, n.d.

Floren, Terese M. "Women Firefighters: The Chief's Role." *Fire Chief Magazine,* May 1981, p. 48.

Lipkin, Harriet. "Affirmative Action — A Solution to Discrimination." *The International Fire Chief,* July 1982, p. 10.

Shindell, Anne B. "Consent Decrees and Affirmative Action Orders — How to Handle Layoffs to Avoid Fire Department Liability." *The International Fire Chief,* July 1982, p. 12.

CHAPTER 3 ENDNOTES

1. "Fire Suppression Rating Schedule." New York: Insurance Services Office, 1980, p. 23.
2. Clark, W. E. *Fire Fighting Principles and Practices.* New York: Dun-Donnelly, 1974.
3. "Manpower Analysis." Dallas Fire Department, 1969.
4. Walker, W., J. Chaiken, and E. Ingnal. *Fire Department Deployment Analysis.* New York: North-Holland, 1979, p. 55.
5. "Fire Suppression Rating Schedule," p. 12.
6. "Fire Suppression Crew Size Study." New York: Centaur Associates, Inc. and U.S. Fire Administration, 1980.
7. Clark, W.E. "Training for Fire Loss Management." A report to the National Fire Academy, 1976.
8. "Municipal Fire Service Workbook." Washington, D. C.: National Science Foundation, 1977.
9. Walker, Chaiken, and Ingnal, p. 161.
10. McKinnon, G. and J. Tower, ed. *Fire Protection Handbook,* 14th edition, Boston: National Fire Protection Association, 1976.

ADDITIONAL REFERENCES

Niemczewski, Christopher. "Fire Suppression Crew Size." *The International Fire Chief,* January 1983, p. 12.

Waters, John M. "Fiscal Crisis in the Fire Service." *Fire Chief Magazine,* October 1981, p. 51.

CHAPTER 4 REFERENCES

Bahme, Charles W. *Fireman's Law Book*. Boston: National Fire Protection Association, 1967.

Bahme, Charles W. *Fire Service and the Law*. Boston: National Fire Protection Association, 1967.

Rappaport, Allen and J.L. Eatman. "Service Delivery During Strikes." *Fire Service Today,* July 1982, p. 11.

Rosenbauer, D.L. Introduction to Fire Protection. Boston: National Fire Protection Association, 1978.

Rhyne, Charles S. *Police and Firefighters: The Law of Municipal Personnel Regulations*. Washington, D.C.: The Law of Local Government Operations Project, Government Law Series, 1982.

Rynecki, Steven B., Douglas A. Cairns, and Donald J. Cairns. *Firefighter Collective Bargaining Agreements: A National Management Survey*. Washington, D.C.: National League of Cities, 1979.

CHAPTER 5 REFERENCES

Bowen, John E. "Aid Your Department by Learning the Basics of Statistics." *American Fire Journal,* February 1984, p. 26.

Burns, H.W. "File System for Records, Documents." *Fire Engineering,* September 1980, p. 130.

Heinen, Larry, Susan Isman, and Warren E. Isman. "Personal Computers for the Fire Service." *The International Fire Chief,* June 1984, p. 12.

Isman, Susan, Larry Elliott, and Larry Heinen. "Taking Perplexity, Complexity Out of Computers." *American Fire Journal,* September 1984, p. 20.

Jackson, Durward P. "Information Technology in the Fire Service." *The International Fire Chief,* February 1983, p. 11.

Kelly, G. Morgan. "Kentucky's Regional Data Centers." *The International Fire Chief,* June 1984, p. 16.

NFPA Standard 901, *Uniform Coding for Fire Protection*. Quincy: National Fire Protection Association, 1976.

Sickle, Edward B. "Developing a Key Factor Reporting System." *Fire Chief Magazine,* June 1983, p. 38.

Toregas, Costis and Robert E. Baumgardner. "A Management Tool for Eluding the Budget Axe." *The International Fire Chief,* February 1983, p. 19.

Uniform Records and Reporting System for Fire Departments. North Carolina: League of Municipalities, n.d.

CHAPTER 6 REFERENCES

Canick, Paul M. "What the Fire Chief of the 1970s Should Know About... PPBS." *Fire Chief Magazine,* January 1970, p. 35.

Chaney, Bryon R. "Municipal Fire Insurance: A Potential for Minimizing Fire Loss." *Fire Chief Magazine,* August 1980, p. 87.

Didactic Systems, Inc. *Management in the Fire Service.* Quincy: National Fire Protection Association, 1977.

Estepp, M.H. "Charging for Ambulance Service is a Step Toward Securing Otherwise Unavailable Enhancement Revenue." *The International Fire Chief,* April 1984, p. 62.

"Fire Preventers Seek 'Private' Money to Offset Lack of State's Funds." *Western Fire Journal,* May 1984, p. 14.

Frazier, David R. "Service Fee Plan Draws Fire From Idaho Churches." *Fire Engineering,* July 1974, p. 41.

Gratz, David B. "Understanding the Budget, Parts I and II." *Fire Command!,* January, February 1971, pp. 12, 18.

Hill, Dale L. "Fire Service Fee." *Fire Command,* August 1979, p. 26.

Hunt, James W. "Taxpayer Revolt Sparks Need for New Fire Service Philosophy." *Fire Chief Magazine,* September 1978, p. 33.

Managing Fire Services. Washington, D.C.: International City Management Association, 1979.

"Municipal Fire Insurance: An Alternative to Private Fire Indemnity at Public Expense in Fire Prevention and Suppression." Berkeley: Institute for Local Self Government, 1977.

The Municipal Yearbook. Washington D.C.: International City Management Association, 1981.

Taliaferro, Julian H. "ZBB — A Logical Look at Budgets." *Fire Engineering,* July 1977, p. 28.

Ward William A. "Zero-Based Budgeting in Fire Service Management." *The International Fire Chief,* May 1979, p. 19.

Ward, William A. "Cutback Management: Learning to Manage More Effectively with Fewer Resources." *The International Fire Chief,* February 1982, p. 13.

CHAPTER 7 REFERENCES

Adams, Rich. "The National Registry of Emergency Medical Technicians." *Firehouse,* April 1984, p. 32.

Chatterton, Howard A. "Public Education Cuts Need for EMS Response." *Fire Chief,* November 1980, p. 34.

"The Dallas EMS DATA System." *Fire Chief Magazine,* November 1980, p. 37.

Fuller, Gregory M. and Einon H. Plummer. "EMS Operations: The Other Side of the Hill." *The International Fire Chief,* April 1984, p. 58.

Little, Lt. Willa K. "Continuing Education for Paramedics: A Necessity." *Fire Chief Magazine.* November 1980, p. 42.

Markman, Howard M. "Managing EMS Liability." *The International Fire Chief,* April 1984, p. 54.

McGlown, K. Joanne. "USFA Opens EMS Resource Center." *Fire Chief Magazine,* November 1980, p. 41.

Page, James O. "Understanding the Fire Service." *JEMS,* June 1984, p. 30.

Page, James O. "Trends in Fire Service EMS." *Fire Service Today,* February 1983, p. 14.

"Paramedic Burnout." *Fire Chief Magazine,* November 1980, p. 27.

Seaver, Jon F. "EMT Insight: The Disturbed and Unruly City." *JEMS,* June 1984, p. 72.

Smith, Bradley H. "Profiles: Defining Fire Service EMS." *JEMS,* June 1984, p. 72.

St. John, Dorothea R. "EMS — Do We Really Need Paramedics?" *Fire Command,* August 1984, p. 28.

CHAPTER 8 REFERENCES

Adams, Rich. "Dallas Dispatch Breakdown." *Firehouse,* May 1984, p. 46.

Bercovici, Martin W. and C. Douglas Jarrett. "The Federal Communications Commission." *Fire Engineering,* March 1983, p. 33.

"Computer Dispatches Preplanning Information." *Fire Chief Magazine,* January 1984, p. 48.

Dektar, Cliff. "Radio Communications Improved Between Paramedics and Hospitals." *Fire Engineering,* March 1983, p. 32.

Demers, David P. "Preplan for Communications Failure." *Fire Command,* March 1980, p. 25.

Ditzel, Paul. "The New Look in Fire Service Communications." *The International Fire Chief,* October 1984, p. 20.

Furey, Barry. "Assessing Radio Procedures." *Firehouse,* September 1983, p. 28.

Furey, Barry. "1984: Getting the Message." *Firehouse,* May 1984, p. 54.

Garnett, Kitty and Wayne. "Non-English-Speaking Communities: Language Forms Barrier to Saving Lives." *American Fire Journal,* August 1984, p. 30.

Hildebrand, Joanne Fish. "Stress Research." Parts I-IV, *Fire Command,* May, June, July, August 1984.

Lamm, W.H. "Techniques Revealed: Dispatching in Emergencies." *Western Fire Journal,* January 1980, p. 16.

Lowden, Thaddeus T. "One Department's Experience: Choosing a Microcomputer." *Fire Command,* February 1984, p. 24.

Press, Wayne. "Keys to Operating a Regional Dispatch Center." *Fire Chief Magazine,* October 1981, p. 58.

Quintanilla, Guadalupe. "Speaking Their Language — Houston Fire Fighters Learn to Work with the Hispanic Citizenry." *The International Fire Chief,* October 1984, p. 23.

Robinson, Vicki. "Call Screening Targets False Emergencies." *The International Fire Chief,* June 1981, p. 16.

Routley, J. Gordon. "C-A-D Phoenix Style." *Fire Engineering,* July 1983, p. 61.

Severino, Frank J. "A Case for Dispatcher Training." *Fire Chief,* February 1984, p. 37.

Simpson, Bob D. "An Inside Look at Fire Service Communications." *Fire Engineering,* March 1983, p. 28.

Simpson, Bob D. "Fire Radio Frequency Coordination." *Fire Engineering,* March 1983, p. 33.

Weiman, Ellen and Marilyn Rooth. "Call Screening." *Firehouse,* May 1983, p. 60.

CHAPTER 9 REFERENCES

Coleman, Ronny J. "Preventive Medicine for Fire Protection." *Fire Chief Magazine,* May 1984, p. 39.

Davis, Paul O., Robert J. Biersner, R. James Barnard, and James Schamadan. "How Fit are Fire Fighters?" *Fire Service Today,* December 1982, p. 11.

Devlin, Frank. "Fire Fighter Fitness — An Often Overlooked Area of Maintenance." *Fire Engineering,* January 1984, p. 32.

Howe, Phil. "Physical Fitness Training Works." *The International Fire Chief,* April 1983, p. 78.

Jacobs, Donald T. *Physical Fitness and the Fire Service.* Boston: National Fire Protection Association, 1976.

Jenaway, William F. "The Fire Chief as Loss Control Manager." *Fire Chief Magazine,* June 1983, p. 36.

King, Kelly and Susan Faerber. "USFA Studies Fire Fighter Injury and Illness — a Followup." *The International Fire Chief,* April 1983, p. 24.

Jacobs, Donald T. *Physical Fitness and the Fire Service.* Boston: National Fire Protection Association, 1976.

CHAPTER 10 REFERENCES

Bruno, Hal. "Unified Command of Fire Legislation." *Firehouse,* February 1983, p. 8.

Coleman, Ronny J. " 'They' are at it Again." *The California Fireman,* April 1982, p. 10.

Dawson, Thomas W. "Managing the Bureaucracy." *The International Fire Chief,* February 1981, p. 18.

Dye, Thomas R. *Politics in States and Communities.* Englewood Cliffs: Prentice-Hall, Inc., 1981.

Keefe, Thomas P. "Effectively Marketing Your Dept. to the Public." *American Fire Journal,* August 1984, p. 16.

Lineberry, Robert and Ira Sharkansky. *Urban Politics and Public Policy,* 2nd ed. New York: Harper & Row, 1974.

McGibeny, Michael D. and Colin A. Campbell. "Landmark Sprinkler Legislation in Florida." *The International Fire Chief,* March 1984, p. 46.

Salisbury, Robert H. ed. *Interest Group Politics in America.* New York: Harper and Row, 1970.

Stratham, Robert R. "A Guide to Effective Lobbying." *Association and Society Manager,* October, November 1984, p. 22.

Stewart, Robert A. "The Local Fire Chief and the Federal Government: Working Through Mutual Expectations." *The International Fire Chief,* April 1980, p. 12.

Truman, David B. *The Government Process,* 2nd ed. New York: Knopf, 1971.

Ward, William A. and James B. Thurman. "How to Work with Local Government to Get Your Budget Approved." *The International Fire Chief,* October 1981, p. 16.

Wilson, James Q. *Political Organizations.* New York: Basic Books, 1973.

Zeigler, L. Harmon and G. Wayne Peak. *Interest Groups in American Society,* 2nd ed. Englewood Cliffs: Prentice-Hall, 1972.

Zisk, Betty H. *Local Interest Politics: A One-Way Street.* Indianapolis: Bobbs & Merrill, 1973.

Index

A

Accident prevention, 177-179
  costs of injuries, 179-181
  domino theory, 177-178
  physical fitness, 183-186
  safety programs, 186-187
Affirmative action, 37-43
  documentation, 41-43
  implementation, 41
  recruitment, 45
  test validity, 43-45
Alarm center, 163-164
  evaluation of, 172-174
  staffing, 164-165
Alarm systems, 153-163
  private alarm systems, 161-163
  public telephone system, 153-155
  radio systems, 157-161
  telegraph systems, 155-156
  telephone fire alarm systems, 156-157
Annual reports, 101-102

B

Budgeting, 111-131
  alternative sources of funds, 128-131
  budgetary process, 124-127
  control, 127-128
Budgeting methods, 113-124
  integrative budgeting system, 120-122
  line-item budget, 113-115
  performance budget, 115-116
  planning-programming budgeting system, 120
  program budget, 116-120
  zero-based budgeting, 122-124

C

Call screening for EMS, 167-168
Collective bargaining, 67-72
Communications systems, 153-174
  alarm centers, 163-164
  alarm systems, 153-163
  computer-aided dispatch, 169-172
  evaluating the communications system, 172-174
Community affairs,
  fire service involvement in, 193-196
Computer-aided dispatch, 169-172

D

Disaster preparedness planning, 17-22, 29-31
Dispatching procedures, 165-168
  retrieval of pre-fire planning information, 100

E

Emergency medical services, 135-150
  certification of EMS personnel, 143-145
  determining need for EMS, 136-139
  dispatching for, 167-168
  good samaritan laws, 145-146
  levels of EMS, 139-143
  managing an EMS, 146-149
  stress and burnout, 149-150
Equal employment opportunity, 35-37
  recruitment, 45

F

Federal communications commission, 160-161
Fire company staffing, 51-58
  emergency medical services, 58
  factors affecting, 54-57
  first alarm assignment, 55-56
  staffing trends, 51-54
Firefighter deaths, causes of, 183
Funds, alternative sources of, 128-131
  for safety, 179-181

H

Hazardous materials incidents,
  planning for, 22-26
Hiring practices, 35-48
  affirmative action, 37-43
  civilian employees, 47
  employment criteria, 47
  equal employment opportunity, 35-37
  recruitment, 45

I

Information management, 91-108
  information transfers, 100-102
  public relations, 102-108
  record systems, 93-95
  report filing systems, 95-96
  statistical analysis, 96-99
ISO rating schedule, 52-53

J

Job applicants, testing for, 43-45
Job descriptions, 42

L

Labor relations, 63-87
  collective bargaining, 67-72
  development of unions, 63-66
  employee actions, 77-78
  handling an impasse, 72-74
  strike management, 74-86
  strike settlement, 84-87
Leadership role of chief, 192-196
  Increasing visibility of department, 193-196
Legislation, 196-201
  communications regarding, 199-200
  legislative process, 197-198
  local campaigns, 196-197
  state campaigns, 197-201
Lobbying activities, 201
Local politics, 192-197

M

Master planning, 26-29
Mutual aid, 13-16

N

National Interagency Incident Management System (NIIMS), 29-31
News media, relations with, 78, 80, 104-108
  newspaper coverage, 106-108
  relations during a strike, 78-80
  on the fireground, 104-105

P

Planning, 5-32
  elements of planning process, 6-12
  hazardous materials incidents, 22-26
  master planning, 26-29
  need for, 5
  problems, 7
  regional planning, 16
  special project planning, 22-26
  time-reflex planning system, 10
  use of flowchart, 11
Political activity, 191-201
  at the local level, 192-197
  at the state level, 197-201
Public relations
  during a strike, 78-80
  performing public education, 102

R

Radio frequencies, 159-160
Record systems, 93-95
  information applications, 99
Regional planning, 16
Report filing systems, 95
  computer storage, 96

S

Safety, 177-187
  documentation of injuries, 181-183
  economics of, 179-181
  physical fitness, 183-186
State politics, 197-201
Statistical analysis, 96-99
Stress and burnout, management of, 149-150
Strike management, 74-86
  contingency plans, 75-77
  continuation of emergency services, 82-83
  internal procedures, 80-84
  public relations and communications, 78-80
  settling the strike, 84-87

T

Testing for job applicants, 43-45

U

Unionization, 63-66

IFSTA MATERIALS

FIRE SERVICE ORIENTATION & INDOCTRINATION
History, traditions, and organization of the fire service; operation of the fire department, and responsibilities and duties of firefighters; fire department companies and their functions; glossary of fire service terms.

FIRE SERVICE FIRST AID PRACTICES
Brief explanations of the nervous, skeletal, muscular, abdominal, digestive, and genitourinary systems; injuries and treatment relating to each system; bleeding control and bandaging; artificial respiration, cardiopulmonary resuscitation (CPR), shock, poisoning, and emergencies caused by heat and cold; fractures, sprains, and dislocations; emergency childbirth; short-distance transfer of patients; ambulances; conducting a primary and secondary survey.

ESSENTIALS OF FIRE FIGHTING
This manual was prepared to meet the objectives set forth in levels I and II of NFPA, *Fire Fighter Professional Qualifications, 1981*. Included in the manual are the basics of fire behavior, extinguishers, ropes and knots, self-contained breathing apparatus, ladders, forcible entry, rescue, water supply, fire streams, hose, ventilation, salvage and overhaul, fire cause determination, fire suppression techniques, communications, sprinkler systems, and fire inspection.

IFSTA'S 500 COMPETENCIES FOR FIREFIGHTER CERTIFICATION
This manual identifies the competencies that must be achieved for certification as a firefighter for levels I and II. The text also identifies what the instructor needs to give the student, NFPA standards, and has space to record the student's score, local standards, and the instructor's initials.

FIRE SERVICE GROUND LADDER PRACTICES
Various terms applied to ladders; types, construction, maintenance, and testing of fire service ground ladders; detailed information on handling ground ladders and special tasks related to them.

FIRE HOSE PRACTICES
Construction, care, and testing of hose and various fire hose accessories; preparation and manipulation of hose for rolls, folds, connections, carries, drags, and special operations; loads and layouts for fire hose.

SALVAGE AND OVERHAUL PRACTICES
Planning and preparing for salvage operations, care and preparation of equipment, methods of spreading and folding salvage covers, most effective way to handle water runoff, value of proper overhaul and equipment needed, and recognizing and preserving arson evidence.

FORCIBLE ENTRY, ROPE AND PORTABLE EXTINGUISHER PRACTICES
Types of forcible entry tools and general building construction; use of tools in opening doors, windows, roofs, floors, walls, partitions and ceilings; types, uses, and care of ropes, knots, and portable fire extinguishers.

SELF-CONTAINED BREATHING APPARATUS
This manual is the most comprehensive text available on self-contained breathing apparatus. Beginning with the history of breathing apparatus and the reasons they are needed, to how to use them, including maintenance and care, the firefighter is taken step by step with the aid of programmed-learning questions and answers throughout to complete knowledge of the subject. The donning, operation, and care of all types of breathing apparatus are covered in depth, as are training in SCBA use, breathing-air purification, and recharging cylinders. There are also special chapters on emergency escape procedures and interior search and rescue.

FIRE VENTILATION PRACTICES
Objectives and advantages of ventilation; requirements for burning, flammable liquid characteristics and products of combustion; phases of burning, backdrafts, and the transmission of heat; construction features to be considered; the ventilation process including evaluating and size up is discussed at length.

FIRE SERVICE RESCUE PRACTICES
IFSTA's *Rescue* has been enlarged and brought up-to-date. Sections include water and ice rescue, trenching, cave rescue, rigging, search-and-rescue techniques for inside structures and outside, and taking command at an incident. Also included are vehicle extrication and a complete section on rescue tools. The book covers all the information called for by the rescue sections of NFPA 1001 for Fire Fighter I, II, and III, and is profusely illustrated.

THE FIRE DEPARTMENT COMPANY OFFICER
This manual focuses on the basic principles of fire department organization, working relationships, and personnel management. For the firefighter aspiring to become a company officer and the company officer who wishes to improve management skills this manual will be invaluable. This manual will help individuals develop and improve the necessary traits to effectively manage the fire company.

FIRE CAUSE DETERMINATION
Covers need for determination, finding origin and cause, documenting evidence, interviewing witnesses, courtroom demeanor, and more. Ideal text for company officers, firefighters, inspectors, investigators, insurance and industrial personnel.

PRIVATE FIRE PROTECTION & DETECTION
Automatic sprinkler systems, special extinguishing systems, standpipes, detection and alarm systems. Includes how to test sprinkler systems for the firefighter to meet NFPA 1001.

INDUSTRIAL FIRE PROTECTION
Devastating fires in industrial plants do occur at a rate of 145 fires every day. *Industrial Fire Protection* is the single source document designed for training and managing industrial fire brigades.

This text is a must for all industrial sites, large and small, to meet the requirements of the Occupational Safety and Health Administration's (OSHA) regulation 29 CFR part 1910, Subpart L, concerning incipient industrial fire fighting.

HAZ MAT RESPONSE TEAM LEAK & SPILL GUIDE
A brief, practical treatise that reviews operations at spills and leaks. Sample S.O.P. and command recommendations along with a decontamination guide.

FIRE SERVICE INSTRUCTOR
Characteristics of good instructor; determining training requirements and what to teach; types, principles, and procedures for teaching and learning; training aids and devices; conference leadership.

PUBLIC FIRE EDUCATION
A valuable contribution to your community's fire safety. Includes public fire education planning, target audiences, seasonal fire problems, smoke detectors, working with the media, burn injuries, and resource exchange.

FIRE PREVENTION AND INSPECTION PRACTICES
Fire prevention bureau and inspecting agencies; fire hazards and causes; prevention and inspection techniques; building construction, occupancy, and fire load; special-purpose inspections; inspection forms and checklists, along with reference sources; maps and symbols; records and reports.

WATER SUPPLIES FOR FIRE PROTECTION
Importance, basic components, adequacy, reliability, and carrying capacity of water systems; specifications, installation, maintenance, and distribution of fire hydrants; flow test and control valves; sprinkler and standpipe systems.

FIRE APPARATUS PRACTICES
Various types of fire apparatus classified by functions; driving and operating apparatus including pumpers, aerial ladders, and elevating platforms; maintenance and testing of apparatus.

FIRE STREAM PRACTICES
Characteristics, requirements, and principles of fire streams; developing, computing, and applying various types of streams to operational situations; formulas for application of hydraulics; actions and reactions created by applying streams under different circumstances.

FIRE PROTECTION ADMINISTRATION
A reprint of the Illinois Department of Commerce and Community Affairs publication. A manual for trustees, municipal officials, and fire chiefs of fire districts and small communities. Subjects covered include officials' duties and responsibilities, organization and management, personnel management and training, budgeting and finance, annexation and disconnection.

FIREFIGHTER SAFETY
Basic concepts and philosophy of accident prevention; essentials of a safety program and training for safety; station house facility safety; hazards en route and at the emergency scene; personal protective equipment; special hazards, including

chemicals, electricity, and radioactive materials; inspection safety; health considerations.

FIRE PROBLEMS IN HIGH-RISE BUILDINGS
Locating, confining, and extinguishing fires; heat, smoke, fire gases, and life hazards; exposures, water supplies and communications; pre-fire planning, ventilation, salvage and overhaul; smokeproof stairways and problems of building design and maintenance; tactical checklist.

AIRCRAFT FIRE PROTECTION AND RESCUE PROCEDURES
Aircraft types, engines, and systems, conventional and specialized fire fighting apparatus, tools, clothing, extinguishing agents, dangerous materials, communications, pre-fire planning, and airfield operations.

GROUND COVER FIRE FIGHTING PRACTICES
Ground cover fire apparatus, equipment, extinguishing agents, and fireground safety; organization and planning for ground cover fire; authority, jurisdiction, and mutual aid, techniques and procedures used for combating ground cover fire.

FIRE SERVICE PRACTICES FOR VOLUNTEER AND SMALL COMMUNITY FIRE DEPARTMENTS
A general overview of material covered in detail in *Forcible Entry, Ladders, Hose, Salvage and Overhaul, Fire Streams, Apparatus, Ventilation, Rescue, Inspection,* and *Self-Contained Breathing Apparatus,* and *Public Fire Education.*

INSTRUCTOR GUIDE SETS
Available for *Forcible Entry, Hose, Salvage and Overhaul, Fire Streams, Apparatus, Ventilation, Rescue, First Aid, Inspection, Aircraft,* and for the slide program *Fire Department Support of Automatic Sprinkler Systems.* Basic lesson plan, tips for instructor, references.

TRANSPARENCIES
Multicolored overhead transparencies to augment *Essentials of Fire Fighting* are now available. Since costs and availability vary with different chapters, contact IFSTA Headquarters for details. Units available:

Fire Behavior; Portable Extinguishers; Ropes and Knots; Hose Tools and Appliances; Handling Hose; Handling Ground Ladders; Ventilation; Fire Streams; Ladder Carries and Raises; Forcible Entry; Salvage and Overhaul; Prevention and Identification; Ground Cover Fires; Communications; Water Supply; Automatic Sprinkler Systems; Rescue; Protective Breathing Apparatus.

FIREFIGHTER VIDEOTAPE SERIES
Video programs for reinforcement of basic skills and knowledge on a variety of fire fighting topics. Excellent for use with *Essentials* or *Volunteer* to review and emphasize subjects. Titles available: The Anatomy and Behavior of Fire, Fire Safety, Protective Breathing Apparatus, Fire Hose and Nozzles — Part 1, Fire Hose and Nozzles — Part 2, Ventilation, Sprinklers, Ladders, Forcible Entry, Rescue, Ropes and Knots, Salvage, Fire Alarm and Communications, General Qualifications, First Aid, Inspection — Part 1, and Inspection — Part 2.

SLIDES
2-inch by 2-inch slides that can be used in any 35 mm slide projector; supplements to respective manuals and sprinkler guide sets.

Sprinklers
Module 1: Introduction to Automatic Sprinkler Protection
Module 2: Types of Sprinkler Systems
Module 3: Maintenance and Inspection of Sprinkler Systems
Module 4: Components of Water Supply Systems
Module 5: Testing and Analysis of Water Supply Systems
Module 6: Factors Affecting the Adequacy of Sprinkler and Water Systems

Smoke Detectors Can Save Your Life
Matches Aren't For Children
Public Relations for the Fire Service
Public Fire Education Specialist (Slide/Tape)
Salvage*

*The complete package consists of the slides, instructor's manual, and instructor's guide sets.

MANUAL HOLDER
The fast, efficient way to organize your IFSTA manuals. Those attractive heavy-duty vinyl holders have specially designed side panels that allow easy access to all of your IFSTA manuals. Manual holders stand unsupported and will hold up to eight manuals.

IFSTA BINDERS
Heavy-duty three ring binders that will allow you to organize and protect your IFSTA manuals. Available in two sizes, $1\frac{1}{2}$ inch and 3 inch.

GUIDE SHEET BINDERS
Free with purchase of complete guide set. Binders also available separately.

WATER FLOW TEST SUMMARY SHEETS
50 summary sheets and instructions on how to use; logarithmic scale to simplify the process of determining the available water in an area.

PERSONNEL RECORD FOLDERS
Personnel record folders should be used by the training officer for each member of the department. Such data as IFSTA training, technical training (seminars), and college work can be recorded in the file, along with other valuable information. Letter size or legal size.

Ship to: _____ **Date** _____

_____
Name   Customer Number

_____
Organization   Phone

_____
Address

_____
City   State   Zip

Send to
Fire Protection Publications
Oklahoma State University
Stillwater, Oklahoma 74078
(405) 624-5723
Or Contact Your Local Distributor

ORDER FORM

IFSTA MANUALS
WRITE THE NUMBER OF COPIES OF EACH MANUAL NEXT TO ITS TITLE

	No. Of Each		No. of Each		No. of Each		No. of Each
Indoctrination	____	Chief Officer	____	Fire Protection Administration	____	Matches Aren't For Children	____
First Aid	____	Fire Cause Determination	____	Safety	____	Public Relations for the Fire Service	____
Essentials	____	Private Fire Protection	____	Aircraft	____		
500 Competencies for Essentials	____	Industrial Fire Protection	____	High-Rise	____		
Ladders	____	Haz Mat Leak and Spill Guide	____	Volunteer	____		
Hose	____	Instructor	____	Ground Cover	____		
Salvage and Overhaul	____	Public Fire Education	____	Manual Binder	____		
Forcible Entry	____	Fire Prevention/ Inspection	____	**SLIDES**			
Self-Contained Breathing Apparatus	____	Water Supplies	____	Ladder	____		
Ventilation	____	Fire Streams	____	Salvage	____		
Rescue	____	Apparatus Practices	____	Public Fire Education Specialists	____		
Company Officer	____			Smoke Detectors Can Save Your Life	____		

VISUAL AIDS
Multicolored overhead transparencies to augment each chapter of *Essentials of Fire Fighting* are now available. Also available are slide programs for each of the major fire pump manufacturers and slides for sprinkler systems. A new addition, The firefighter videotape series is available for both Beta and VHS, and ¾" formats. Since costs and availability vary with different sets, contact Fire Protection Publications for details.

OTHER MANUALS AND MATERIALS MAY BE ORDERED BELOW:

QUANTITY	TITLE	LIST PRICE	TOTAL

All Foreign Orders must be prepaid in U.S. currency and include 20% shipping and handling charges.

Obtain postage and prices from current IFSTA Catalog or they will be inserted by Customer Services.

Note: Payment with your order saves you postage and handling charges when ordering from Fire Protection Publications.

Payment Enclosed ☐   Bill Me Later ☐

Allow 4 to 6 weeks for delivery.

SUBTOTAL $ _____
Discount, if applicable $ _____
Postage and Handling, if applicable $ _____
TOTAL $ _____

FOR ORDERS
TOLL FREE NUMBER — 800-654-4055

Oklahoma, Hawaii, and Alaska call collect.

COMMENT SHEET   **CHIEF OFFICER**

DATE _____ NAME _____

ADDRESS _____

ORGANIZATION REPRESENTED _____

CHAPTER TITLE _____ NUMBER _____

SECTION/PARAGRAPH/FIGURE _____ PAGE _____

1. Proposal (include proposed wording, or identification of wording to be deleted), OR PROPOSED FIGURE:

2. Statement of Problem and Substantiation for Proposal:

RETURN TO: IFSTA Editor   SIGNATURE _____
                     Fire Protection Publications
                     Oklahoma State University
                     Stillwater, OK 74078

Use this sheet to make any suggestions, recommendations, or comments. We need your input to make the manuals the most up to date as possible. Your help is appreciated. Use additional pages if necessary.